FUNDAMENTALS OF THE CHEMISTRY
AND
APPLICATION OF DYES

Fundamentals of the Chemistry and Application of Dyes

P. Rys
and
H. Zollinger

*Eidgenössische Technische Hochschule (E.T.H.)
Zurich*

WILEY–INTERSCIENCE
a division of John Wiley & Sons Ltd
London   New York   Sydney   Toronto

Copyright © 1972 John Wiley & Sons Ltd. All Rights Reserved. No part of this publication may be reproduced, stored in a retrieval system, or transmitted, in any form or by any means, electronic, mechanical photocopying, recording or otherwise, without the prior written permission of the Copyright owner.

Library of Congress catalog card number 78-37108

ISBN 0 471 74795-5

Printed in Northern Ireland at The Universities Press, Belfast

Preface

The evolution of the chemistry of dyes began in 1856 with the discovery and industrial production of Mauvein by W. H. Perkin. The essential basis of modern chemistry evolved at the same time (periodic system, quadrivalence of carbon, the structure of benzene). Thus the chemistry of dyes became that field of technical organic chemistry in which scientific methods were first applied. Other branches did not follow until the end of the 19th century.

It seems to us that this historical background nowadays exerts an inhibiting effect on dyestuff chemistry; it is frequently regarded as closed, conservative and not readily amenable to new discoveries. For this reason teaching institutions pay little attention to it.

However in a number of the more highly developed countries dyestuff chemistry is the basis of an important section of the chemical industry, which is actively engaged on research. Therefore, on the one hand, chemists who enter the dyestuff industry should be specially trained for work in this field but it is essential, on the other hand, that dye chemists always view their work in relation to the present state of general chemistry.

The present introductory work will attempt to meet both of these requirements; we shall provide the beginner who has read little specialized dyestuff chemistry at technical college with a review that covers all classes of dyes, but is confined to essential and characteristic structural features. We have resisted the temptation to enumerate as many commercial dyes as possible; this is the purpose of the Colour Index! We intend to demonstrate how the domain of dyes is connected with the fundamentals of modern inorganic, organic and physical chemistry, and to show that the enlightened dye chemist also requires this knowledge in his sphere of work.

We would also like to show the familiar field in a new light to the experienced, conventionally trained, dyestuff chemist, and induce him to devote his attention to unsolved problems. The more theoretical chapters 2, 3 and 11 are meant as an introduction to the physical and chemical aspects: naturally, they offer little that is new to the true physical chemist.

The only book written after the second World War which considers both the basic principles of dyes in textile chemistry as well as their nontextile applications is the classical work of K. Venkataraman (1952–53, supplementary volumes 1970–71). As we feel that this will be the most productive approach in the future, a large portion of this book will deal also with nontextile aspects of dyestuff chemistry.

The book is based on lectures which both of us have given at the Swiss Federal Institute of Technology (E.T.H.) Experience has taught us that the only way of gaining new friends for dyestuff chemistry is to treat dyes, not as a special section, but as an integrated component of organic and physical chemistry.

Several colleagues have given us valuable hints on planning this book. We are particularly indebted to Dr. H. H. Bosshard (Ciba-Geigy Ltd.), Dr. E. Ganz (Ciba-Geigy Ltd.), Prof. E. Heilbronner (University of Basle, formerly E.T.H.), Prof. W. Jenny (Ciba-Geigy Ltd.), Dr. H. Klessinger (University of Göttingen), Prof. W. Lüttke (University of Göttingen), Dr. H. E. Nursten (University of Leeds), Dr. R. Orwell (University of New South Wales) and Dr. P. B. Weisz (Mobil Oil Corp., Princeton, U.S.A.) for their advice and discussions.

We hope that this small book will help to arouse and increase the interest of industrial chemists, college lecturers and students in dye chemistry.

Zurich, August 1970   Paul Rys
                                                        Heinrich Zollinger

Contents

1. Introduction **1**

    1.1. History and industrial production of dyes 1
    1.2. The classification of dyes 3
    1.3. Literature 4

2. The Basis of Colour in Organic Compounds **5**

    2.1. Relation between chemical constitution and light absorption . . 5
    2.2. Colorimetry 15
    2.3. Literature 21

3. The Fundamentals of the Reaction Mechanisms of Dyestuff Syntheses. **23**

    3.1. Introduction 23
    3.2. Preequilibria in aromatic substitution. 24
    3.3. Mechanism of S_E2 reactions 29
    3.4. Nucleophilic aromatic substitutions 33
    3.5. The quinone-hydroquinone redox system 34
    3.6. Equilibria in the formation of metal complexes 37
    3.7. Literature 40

4. Azo Dyes **42**

    4.1. Principles underlying the preparation of azo dyes . . . 42
        4.1.1. General methods 42
        4.1.2. Diazotization 44
        4.1.3. Azo coupling 45
    4.2. Application in dyeing 48
        4.2.1. Anionic monoazo dyes 48
        4.2.2. Disperse azo dyes 51
        4.2.3. Azoic dyes 52
        4.2.4. Complex-forming monoazo dyes 54
        4.2.5. Direct dyes 58
        4.2.6. Reactive dyes 62
    4.3. Literature 70

5. Nitro and Nitroso Dyes 72

6. Polymethine Dyes 74

6.1. General discussion and structures 74
6.2. The principles underlying the preparation of polymethine dyes . 79
6.3. Application of polymethine dyes in dyeing 83
6.4. The application of polymethine dyes in photography . . . 84
6.5. The use of polymethine dyes in colour photography . . . 85
6.6. Literature 91

7. Aza[18]annulenes 92

7.1. General remarks and structures 92
7.2. Principles of preparation 95
7.3. Technical application of aza[18]annulene dyes 97
7.4. Literature 98

8. Di- and Triarylcarbonium Dyes and their Aza Analogues . . 99

8.1. General remarks and structures 99
8.2. Principles of preparation 101
8.3. Application of di- and triarylcarbonium dyes in dyeing . . 106
8.4. The application of di- and triarylcarbonium dyes as indicators . 107
8.5. Literature 109

9. Sulphur Dyes 110

9.1. General remarks 110
9.2. Structures and principles of preparation 110
9.3. Application of sulphur dyes 112
9.4. Literature 113

10. Carbonyl Dyes 114

10.1. Indigo and its derivatives 114
   10.1.1. General remarks and structures 114
   10.1.2. Methods of synthesis used in indigo chemistry . . 119
10.2. Anthraquinone substitution products 124
   10.2.1. General remarks 124
   10.2.2. Structures and principles of preparation . . . 125

Contents

10.3. Higher anellated carbonyl compounds	140
10.3.1. General remarks	140
10.3.2. Structures and principles of preparation	141
10.4. Other carbonyl dyes	145
10.5. Application in dyeing processes	149
10.5.1. Ionic carbonyl dyes	149
10.5.2. Disperse dyes	150
10.5.3. Complex-forming carbonyl dyes	151
10.5.4. Reactive dyes	151
10.5.5. Vat dyes	151
10.5.6. Leuco sulphuric ester dyes	156
10.6. Literature	158
11. The Application of Dyes	**160**
11.1. General remarks	160
11.2. Dyeing with organic pigments	160
11.3. Dyeing methods based on equilibrium processes	162
11.3.1. The dyeing system in the state of equilibrium	163
11.3.2. The kinetics of dyeing	174
11.4. Literature	184
Index	**187**

1

Introduction

1.1. HISTORY AND INDUSTRIAL PRODUCTION OF DYES

PREHISTORIC man had already dyed furs, textiles and other objects with natural substances, mainly of vegetable, but also of animal origin. Ancient Egyptian hieroglyphs contain a thorough description of the extraction of natural dyes and their application in dyeing.

Further developments extending over many thousands of years led to rather complicated dyeing processes and high-quality dyeings. Among these the following deserve special mention: Indigo, which was obtained both from dyer's woad, indigenous in Europe, and from *Indigofera tinctoria*, a native plant of Asia; Ancient purple which was extracted from a gland of the purple snail by a process developed by the Phoenicians; Alizarin, on which Turkey Red is based, which was obtained from madder Campeachi wood extract imported from Africa.

Picric acid, which was obtained by Woulfe in 1771 by treating indigo with nitric acid, was subsequently occasionally used for dyeing silk yellow but did not attain any significance. For this reason—really incorrectly—William H. Perkin, not Woulfe, was given credit for having produced the first synthetic organic dye. In 1856, this young, talented, English chemist, working in a suburb of London, succeeded in obtaining by oxidation of a mixture of aniline bases, not quinine, which he had hoped for, but a violet cationic dye which he called Mauveine; initially, the yield was poor. If Perkin is to be regarded as the founder of the synthetic dye industry this is correct in the sense that, on the one hand, with the primitive means then at his disposal he was able to prepare a relatively pure and technically interesting product and, on the other hand, was able to develop its synthesis so that it could be used in large-scale production.

The brilliant violet hue on silk immediately attracted much attention and stimulated other chemists to carry out similar experiments. In this way, in 1859, E. Verguin in Lyon discovered fuchsine, whilst the discovery of diazo compounds by P. Griess in England laid the foundation for the development of the currently largest class of synthetic dyes, namely the azo compounds. The first true azo dye, Bismarck Brown, was developed by Martius in 1863.

After the decisive work by August von Kekulé on the quadrivalence of carbon (1858) and on the constitution of benzene (1865), the way was opened for the planned preparation of purely synthetic dyes, as well as for the artificial production of natural dyes. The first success to be mentioned is the elucidation of the constitution (1868), and the synthesis immediately afterwards, of alizarin (1,2-dihydroxyanthraquinone) by Graebe and Liebermann. The elucidation of the constitution, and the synthesis, of indigo (Adolf von Baeyer, 1870; K. Heumann, 1890) involved research work extending over several decades. The development of indigoid dyes reached a climax at the beginning of the 20th century as a result of the work of G. Engi and of P. Friedländer who synthesized Ciba Blue 4B and thioindigo, respectively. Shortly before the turn of the century, Vidal opened up the field of sulphur dyes, whilst the year 1901 was characterized by the discovery of indanthrone, the first anthraquinonoid vat dye, by R. Bohn.

The natural dyes already included metal complexes (Turkey Red). In this domain, the 20th century yielded important contributions with Neolan dyes (1915), phthalocyanine (1936) and the 1:2-metal complex dyes (Irgalan dyes, 1949). R. Clavel and H. Dreyfuss, in the 1920's, solved the problem of dyeing hydrophobic fibres by means of disperse dyes. The post-war era is characterized by the development of pigments which achieved importance for colouring plastics (e.g. quinacridone, 1958), Cromophtalic dyes (1957), reactive dyes for wool (1951/52: Remalan and Cibalan Brilliant dyes) and especially reactive dyes for cellulose fibres (Procion dyes, 1956). New developments in this field have been reviewed by H. Pfitzner[1] and K. Venkataraman.[2]

Reactive dyes, for example, clearly show the general tendency of dyestuff research to move away from purely empirical syntheses of coloured molecules to the study of the mechanisms of interaction of substrates and dyes with already known chromogens (e.g. azo, anthraquinone and cyclic aza compounds).

The fact that nowadays thousands of dyes of different constitution are commercially available clearly shows that a thorough knowledge of synthetic organic

Table 1.1 A review of dyestuff production 1966/67[3]

	Production			Export	
	10^3 t	10^6 \$	\$/kg	10^3 t	10^6 \$
Western Europe					
German Federal Republic	77·3	278	3·60	46·9	170
Great Britain	38·5	128	3·33	21·4	62
Switzerland	30·9	158	5·35	26·4	141
France	20·7	60	2·93	8·0	23
Italy	13·0	23	1·79	2·1	5
Remainder	17·5	23	—	—	—
U.S.A.	112·1	436	3·89	10·0	35
Japan	48·8	101	2·08	4·6	9
Eastern Europe and China	180	—	—	—	—
Remainder (India, Brazil, Canada, etc.)	30	—	—	—	—
Total	560	—	—	—	—

Introduction

chemistry (leading to the preparation of new dyes, the discovery of new reactive groups, etc.), of the reaction mechanisms (leading to the optimization of manufacturing processes) and of the techniques of applying the results of dye research can ultimately bring success.

The manufacture of dyestuffs is characterized by its concentration, in a few large concerns, in a few countries. Whilst before the first World War almost the whole of the world's requirements of synthetic dyes was manufactured in Germany (apart from Germany, only Switzerland contributed a significant proportion of world production), nowadays manufacture is spread over a larger number of countries. Table 1.1 indicates that Germany is still the largest exporter of dyestuffs, that Switzerland produces particularly high-grade dyes, and that the production of markedly cheap dyes is concentrated in countries like Italy and Japan. Unfortunately, there are no recent reliable statistical data available for Eastern Europe and for China.[3]

1.2. THE CLASSIFICATION OF DYES

Dyes can be grouped in accordance with two different principles:
(a) chemical structure (chemical classification),
(b) dyeing methods and areas of application (colouristic classification).

Chapters 4–10 of this book discuss the most important classes of dyes primarily according to their chemical grouping. Chapter 11 discusses the physical and chemical basis of the colouristic classification, but the more practical aspects of dyeing technology are included at the appropriate places in Chapters 4–10.

A review of the whole field of technical dyes shows that the two classifications overlap, i.e. there is hardly a chemical class of dye which occurs solely in *one* colouristic group, and vice versa. In the same way, some colouristic groups can be applied to two or more substrates, whilst others are specific to a single substrate. Reactive dyes, for example, are classified chemically as coloured compounds having a group capable of forming a covalent bond with a substrate. The coloured parent substance can, in principle, be derived from all classes of Chapters 4–10 but the compounds mainly involved are of the azo, anthraquinone and aza annulene types (sections 4.2.6, 10.5.4 and 7.2, respectively).

When classified according to the dyeing method, they may be anionic, direct or disperse dyes, depending on whether they are intended for use on protein, cellulose or polyamide fibres. Moreover, certain reactive dyes having a particular type of chemical structure can be used for several substrates, whilst others (having the same type of structure) are suitable for only a single substrate.

It is possible to devise a logical method of chemical classification,[4] and the present book is based upon it. However, owing to space limitation, we are unable to discuss it in detail. With some classes of dyes, however, we have abandoned the conventional nomenclature, since it is either wrong (for instance, 'basic' instead of 'cationic' dyes) or insufficiently comprehensive (e.g. phthalocyanine as a sub group of aza annulene structures).

The third edition of the Colour Index (1971)[5] (5 volumes in all) describes all dyes that were being manufactured at the time of going to press. In Part 1, dyes are grouped, according to colour, in the following classes: acid, mordant, basic, disperse, natural dyes and pigments, food, leather, direct, sulphur, vat, reactive, an ingrain section including ingrain dyes, azoic diazo components, azoic coupling components, azoic compositions, oxidation bases, pigment and solvent dyes, optical brighteners, intermediate products, developers and reducing agents. The various classes of dyes are subdivided according to colour, viz. yellow, orange, red, violet, blue, green, brown and black and (for pigments) white.

Part 1 also gives methods of application, usage, the more important fastness properties and other basic data.

Part 2 gives the structural formulae (where known) of the dyes, methods of manufacture and literature references, including patents.

Part 3 includes abbreviations of manufacturers' names, details of fastness tests, a patents index and a commercial names index.

Each dye is given two reference numbers, one relating to its class with respect to dyeing method (e.g. C.I. Vat Blue 4 for indanthrone) and the other to its chemical description in Part 2 (e.g. C.I. 69800 for indanthrone). All the commercial names under which each dye is sold (over 37 for indanthrone e.g. Indanthrene Blue RS, Cibanone Blue RSN, Caledon Blue XRN, Ponsol Blue GZ etc.) are also listed in Part 1.

1.3. LITERATURE

1. H. Pfitzner, in *Ullmann's Encyclopädie der technischen Chemie*, 3rd edition, Urban and Schwarzenberg, München, 1970, Supplementary Volume, p. 217.
2. K. Venkataraman, *The Chemistry of Synthetic Dyes*, Vol. 3-5, Academic Press, 1970/71.
3. L'Industrie chimique 1966/67, Organisation de Coopération et de Développement Economique, Paris, 1968.
4. P. Rys, unpublished work.
5. Colour Index, 3rd edition 1971, Society of Dyers and Colourists, Bradford.

2

The Basis of Colour in Organic Compounds

2.1. RELATION BETWEEN CHEMICAL CONSTITUTION AND LIGHT ABSORPTION

DYES are characterized in accordance with their capacity to absorb the energy of that part of electromagnetic radiation to which the human eye is sensitive. Of the whole spectrum of electromagnetic vibrations, the microwave (wavelength above 500 μm, wave numbers below 20 cm^{-1}), infrared (0·7–500 μm, 15,000–20 cm^{-1}), visible (400–700 nm, 25,000–15,000 cm^{-1}) and ultraviolet (10–400 nm, 10^6–25,000 cm^{-1}) regions are important in chemistry because the molecules, by absorbing energy, are raised to (higher energy) excited states. Absorption of the relatively low-energy microwave and infrared radiations is associated with changes in the rotational and vibrational energy of molecules and atomic groups. However, the higher energy of visible and ultraviolet light (40–70 and 70–3,000 kcal/mol) raises the electrons into excited states (electronic spectra). There is, therefore, no fundamental difference between visible and ultraviolet spectra.†

Naturally, chemists have been occupied with the relation between chemical constitution and colour of organic compounds since the origin of structural chemistry. Several 'chemical' theories have been based on the first, purely empirical, system, which was developed by O. N. Witt[1] in 1876; the authors will briefly deal with these towards the end of this chapter.‡

A full understanding of electronic spectra requires a knowledge of the energy values of the ground and the excited states of the absorbing molecules and ions. According to the Einstein–Bohr frequency condition ($\Delta E = h\nu$), the energy differences ΔE of these states are directly proportional to the frequency (or wave number respectively), i.e. inversely proportional to the wavelength, of the absorbed light. With the aid of W. Heisenberg's quantum mechanics (1925) and E. Schrödinger's wave mechanics (1926) it was possible to obtain information about the energy levels of electrons in atoms and molecules.

In the wave mechanical model for the three-dimensional states of an electron, the electron 'paths' in the original Bohr atom model are replaced by *orbitals*, i.e. the partition functions of single electrons. The Schrödinger equation (2.1) describes the motion of an electron of mass m and potential energy U as a function

---

† Electron transitions are also responsible for bands in the near infrared (up to about 1,000 nm and in isolated cases 1,500 nm).
‡ A. Maccoll[2] has compiled an up-to-date summary.

of the three space coordinates x, y, and z, the total energy E and the eigenfunction ψ. ψ^2 here is a measurable quantity related to the electron density.

$$\frac{\partial^2 \psi}{\partial x^2} + \frac{\partial^2 \psi}{\partial y^2} + \frac{\partial^2 \psi}{\partial z^2} + \frac{8\pi^2 m}{h^2} [E - U(xyz)]\psi = 0 \qquad (2.1)$$

A solution of the Schrödinger equation is possible with atoms which are similar to hydrogen (H, Li$^+$, etc.); however, difficulties are encountered with atoms with more complex electron structure, since the interaction of several electrons has not been precisely elucidated. Although these problems are being intensively studied at the present time, it is still necessary to reduce these poly-electron systems, by means of approximate solutions, to single electron problems.

Of the methods which are used for this purpose, the molecular-orbital method (MO) (Mulliken 1928, Hund and Lennard-Jones) will first be described. Here a linear combination of wave functions (LCAO-MO, linear combination of atomic orbitals) is formed from the atomic orbitals of the valency electrons. With a diatomic molecule, e.g. HD, combination may occur in phase ($\psi' = N'(\psi_H + \psi_D)$), and out of phase ($\psi'' = N''(\psi_H - \psi_D)$) ($N'$ and N'' are normalizing factors). ψ' has a lower energy and ψ'' a higher energy than the electrons of the isolated hydrogen and deuterium atoms. For this reason HD is more stable than H + D; in accordance with the Pauli principle the two electrons of HD have opposite spin and occupy the molecular orbital ψ'; this is termed bonding and ψ'' anti-bonding.

The energy level of ψ' decreases as the extent of the atomic orbital 'overlap' increases. The so-called overlap integral is $S = \int \psi_H \psi_D \, dV$, taken over all space.

The molecule HD can absorb the light energy when one of the two electrons is raised from the bonding (ψ') ground state into the anti-bonding excited state (ψ''). The energy difference between ψ' and ψ'' therefore determines the absorption frequency (or wavelength)—see above. These two states are denoted by σ and σ^* because σ-electrons participate.

Since both electrons in the excited state no longer occupy the same molecular orbital, the restriction imposed by the Pauli principle no longer applies; there are, therefore, two excited states, the singlet and the triplet state, with anti-parallel ($\downarrow\uparrow$) or parallel ($\uparrow\uparrow$) spins of the two electrons, respectively. The triplet state, which is important in respect of phosphorescence, has a lower energy level than the singlet of the same electron configuration.

It is well known, from classical chemical investigations, that benzene is a uniform compound, the properties of which do not correspond to the hypothetical cyclohexatriene. The electron distribution of benzene may be deduced from a molecular-orbital consideration: thus, as with ethylene the two π-orbitals overlap, with the hexagonally-symmetrical benzene there is paired overlapping not only of 3 × 2 adjacent, but of all six, orbitals.

Calculation of the energy levels of the π-electrons, according to the MO method, by E. Hückel[3] has shown that pairs of the second and the third bonding molecular

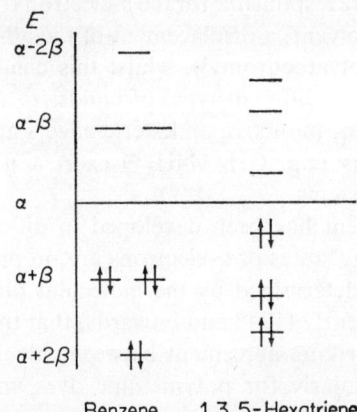

Figure 2.1 Orbitals of benzene and 1,3,5-hexatriene

orbitals as well as the two lower anti-bonding orbitals possess the same energy ('degenerate' MO). The scheme shown in Figure 2.1 illustrates the energy levels of these orbitals calculated by the simplest MO approximation method ('Hückel–MO' or HMO).

Here the so-called Coulomb integral α serves as reference energy level, corresponding approximately to the ionization potential of an electron in a $2p$ atomic orbital. The resonance integral β is a measure of the interaction between two orbitals; it is approximately proportional to the overlap integral S and with aromatic hydrocarbons amounts to about 18–20 kcal/mol. The energy 2β is termed the mesomeric energy, or better the '*delocalization energy*', of benzene.

E. Hückel demonstrated that this particularly strong stabilization occurs only in homo- and heterocyclic compounds which have a total number of $(4n + 2)\pi$-electrons in the perimetric ring system (Hückel rule). As the investigation of annulene systems (Chap. 7) showed, it is of no fundamental importance whether they are mono- or polycyclic systems when $n = 1, 2, 3$, (etc.). The cycloheptatrienylium cation, which E. Hückel in 1931 predicted to be stable (tropylium ion, $C_7H_7^\oplus$), was first prepared by W. E. von Doering in 1956 and it represents one of the neatest proofs of the usefulness of quantum theoretical work in organic chemistry!

Where an aromatic compound carries a group with π-orbitals in a position adjacent to aromatic carbon (e.g. NO_2 or COOH), overlapping of both π-electron systems normally occurs, leading to new bonding and anti-bonding electron states. The same thing occurs on substitution with atoms that have lone pairs of electrons (e.g. OH, NR_2, etc.). These electrons occupy p-orbitals and, for geometrical reasons, they can also overlap with the aromatic π-electrons. Because of the possibility of an interaction of the p-electrons with protic solvents (H-bond), the absorption bands which are responsible for the p-electron transitions exhibit, with increasing polarity of the solvent, a displacement of the absorption bands to shorter wavelengths ('negative solvatochromy'), whilst this change of medium generally has the opposite effect with different types of bands.

When hyperconjugation, inductive and steric effects are neglected, substituents without p- or π-electrons (e.g. CH_3, $NH_3^{\oplus}$) exert a negligible influence on π-orbitals of conjugated systems.

The simple MO treatment has been developed in different ways; with the free electron model[4] the energy states of π-electrons are, in principle, treated as if they vibrate freely in a 'box' determined by the molecular dimensions. H. Kuhn[5] and J. R. Platt[5] showed quite early (1948 and onwards) that this model, using relatively simple means, leads to striking agreement between calculated and experimentally measured spectra, particularly for polymethine dyes and aromatic compounds. The 'self-consistent field' method (SCF)[6] embodies, semi-empirically, the interelectronic interactions, and the Pariser–Parr–Pople method[7] embodies the configuration interactions. In the MIM method (molecules in molecules),[8] with complicated molecules, only those changes (which relate to the known spectra of sub systems) that occur when these systems are coupled to the whole of the molecule are calculated. Here coupling is attributed to the electrostatic interaction between the transition moments occurring in the individual excitation of the subsystems and to an electron transition from one subsystem to another.

The *valence bond method* (VB) uses a starting point different from that of the MO method for the approximate solution of the Schrödinger equation; this is a polyelectron theory which, in contrast to the above-mentioned methods, results in a linear combination of polyelectron functions and not atomic orbitals. These functions can be assigned to different mesomeric structures of the particular molecules. Instead of orbital diagrams (e.g. Figure 2.1), (overall) state diagrams for the different mesomeric structures are obtained. These states are coupled.

A coupling of mesomeric structures which do not exist as such is also termed resonance.

The *qualitative* application of the valence bond method (the so-called resonance theory) has led to many misunderstandings within the last three decades. On the one hand, mesomeric structures were repeatedly assumed to be different structures of the same compound—in the same way as tautomers are discrete isomers ('oscillation between two mesomeric structures')—and on the other hand the mesomeric structures were connected with excited states. Since the term 'resonance' refers to the coupling of two (true) oscillatory systems, it is not as good as the term 'mesomerism', coined by Ingold, which indicates in a better way that the true low energy state lies *between* two (hypothetical) mesomeric structures.

The electron spectrum of a molecule exhibits a multiplicity of absorption bands of widely differing intensity which are partly superimposed. In principle, *one* particular electron transition is assigned to one band. However, these bands are not clearly

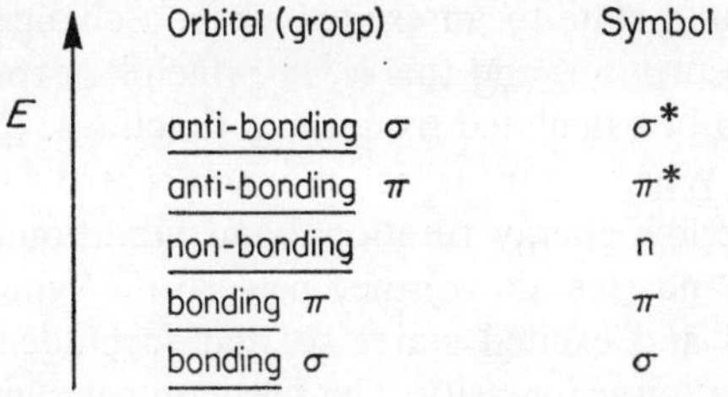

The Basis of Colour in Organic Compounds

Orbital (group)	Symbol
anti-bonding σ	σ^*
anti-bonding π	π^*
non-bonding	n
bonding π	π
bonding σ	σ

Figure 2.2 Scheme showing molecular orbitals

defined since simultaneous changes in the vibrational and rotational states are superimposed on the electron transitions. Figure 2.2 shows, in the simplest form, the essential types of energy levels. From the previous discussion it is clear that with polyatomic molecules more than one of each of these levels is present. As demonstrated in Figure 2.2, a transition of a p-electron from a non-bonding orbital (n) into the anti-bonding π^*-orbital ($n \to \pi^*$-transition) requires light of lowest energy; therefore a $n \to \pi^*$-transition usually occurs at longer wavelengths than a $\pi \to \pi^*$-transition.† In some cases however (e.g. with α,β-unsaturated ketones) a $\pi \to \pi^*$-transition corresponds to the absorption band with the longest wavelength. $\sigma \to \pi^*$- and $\sigma \to \sigma^*$- transitions mainly result in absorption in the far ultraviolet. Not all transitions lead to anti-bonding states; with triphenylmethane dyes, for example, transitions to non-bonding excited states are of importance.

The *intensity* of absorption bands, as well as their wavelength, is of outstanding importance in technical dye chemistry. The molar extinction coefficient ε of the Lambert–Beer law is a measure of the former (2.2).

$$\log \frac{I_0}{I} = E = \varepsilon\, d\, c \qquad (2.2)$$

I_0, I: Intensity of the incident or transmitted light
E: Extinction (or optical density)
d: Layer thickness (cm)
c: Concentration (mol/l)
ε: Molar extinction coefficient (l/mol cm)

For commercial dyes ε has values of from 10^4 (better 2×10^4) to over 10^5 l mol^{-1} cm^{-1}. The area ($\int \varepsilon\, d\tilde{\nu}$) of the absorption bands is important in determining a quantity, the oscillator strength f ($f = 4\cdot 3 \times 10^{-9} \int \varepsilon\, d\tilde{\nu} = \varepsilon_{\max} \Delta\tilde{\nu}_{1/2}$ where $\Delta\tilde{\nu}_{1/2}$ is the width of the band at half height) which is related to the nature of the electronic transition. Strong absorption bands ($f \simeq 1$; $\varepsilon_{\max} = 10^4$–$10^5$) correspond to 'permitted', and weak ones to 'forbidden' electron transitions. When an electron is

† Here we denote excited states with the sign * (some authors also use this sign for anti-bonding orbitals); the form $\pi^* \leftarrow \pi$ (instead of $\pi \to \pi^*$) is also used.

raised from the ground state to an excited state, a change in charge distribution can occur during absorption, and this is, in principle, expressed by the transition integral Q which can be calculated from the ψ functions. Q^2 is proportional to the oscillator strength f. When the transition moment is zero a transition is (symmetry) forbidden. Since the low energy rotational and vibrational oscillations lead to a broadening of the band (see above) they perturb the symmetry of the charge distribution of ground and excited states so that forbidden transitions do occur; however, they have a lower intensity. The selection rules give detailed information as to whether a transition is permitted or forbidden.

The rotational and vibrational processes of the nuclei which are superimposed on an electron transition in the ground and in the excited state are the cause of the

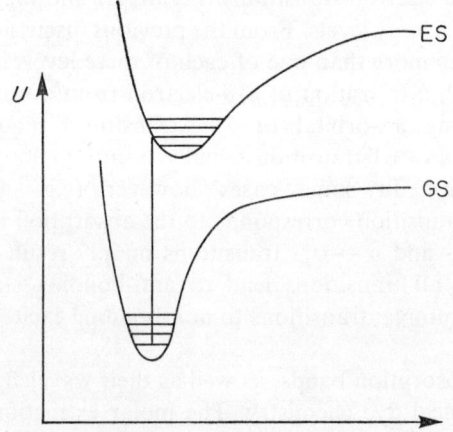

Figure 2.3 Morse curves of the ground (GS) and an excited state (ES) of a diatomic molecule

fine structure of the absorption bands. Their analysis has so far only been possible for diatomic molecules and highly symmetrical polyatomic molecules such as benzene. The potential energy U of a diatomic molecule depends on the internuclear distance, described approximately by the 'Morse curve' (Figure 2.3). GS of the Morse curve corresponds to the ground state, while ES corresponds to that excited state in which the valency electrons are slightly loosened and the internuclear distance is therefore increased. The separate levels relate to the vibrational states of the nuclei. In accordance with the principle laid down by Franck and Condon, the molecular geometry, and the kinetic energy of the oscillations, must remain unchanged during transition since the electron transition is much more rapid (approximately 10^{-14}–10^{-15} s) than the nuclear oscillations (approximately 10^{-10}–10^{-12} s).

This gives rise to a multiplicity of energy variations in an electron transition, with the resulting fine structure of the bands. In accordance with the shape and the position of the Morse curves GS and ES there will be different band structures; the example of Figure 2.3 shows an approximate multi-band spectrum with the most intense bands in the centre. With larger molecules, by comparing the spectra in the vapour and in the solution or the effect of marked increases of temperature, it is found that the fine structure largely disappears owing to the multiple possibilities of intra- and intermolecular interactions. An electron can drop from an excited state (Figure 2.3) without emitting radiation into a lower level of curve ES, so that there is no reversal of spin: it still remains in the singlet state. It may then radiate its energy as fluorescence.†

Table 2.1 Absorption spectrum of azobenzene in hexane[9]

ν	Position of the band λ	ε_{max}	Assignment
22,500 cm^{-1}	450 nm	465 l mol^{-1} cm^{-1}	$n_s \to \pi^*$
31,500 cm^{-1}	330 nm	17,000 l mol^{-1} cm^{-1}	$\pi \to \pi^*$
35,000 cm^{-1}	280 nm	8,200 l mol^{-1} cm^{-1}	$\phi_2 \to \pi^*$
37,500 cm^{-1}	260 nm	3,700 l mol^{-1} cm^{-1}	$n_a \to \phi^*$
40,250 cm^{-1}	240 nm	40,250 l mol^{-1} cm^{-1}	$\pi \to \phi^*$
43,700 cm^{-1}	225 nm	12,400 l mol^{-1} cm^{-1}	$\phi_1 \to \phi_2^*, \phi_2 \to \phi_2^*$

n_s, n_a: Transitions of p-electrons of nitrogen atoms.

π: Transition of the total π-electron system in which the azo group participates (i.e. with transfer character between phenyl and azo groups).

ϕ: Localized excited states of the phenyl ring.

When comparing spectra, displacements of an absorption band to a shorter or longer wave region are commonly termed hypsochromic or bathochromic, respectively.

The authors will now demonstrate, for azobenzene in solution, the assignment, not only of the absorption band of lowest frequency, but also of all other bands of the absorption spectrum in the visible and the near ultraviolet regions (down to 210 nm), in order to show the multiplicity of electron transitions that must be considered, even in a relatively simple organic compound (Table 2.1). It is based on work by Kortüm and Rau as well as other authors.[9]

It will be seen from Table 2.1 that the weak colour of azobenzene is attributable to a transition of one of the lone electrons in the azo nitrogen atoms, but that the strong absorption band at 330 nm is due to a $\pi \to \pi^*$-transition. When a dimethylamino group is introduced in the 4-position of azobenzene, a *visually* significant intensification of the colour occurs and a 'dye' in the technical sense is produced, Dimethyl Yellow, C.I. Solvent Yellow 2 ($\lambda_{max} = 408$ nm, $\varepsilon = 27,500$ l mol^{-1} cm^{-1} in 95 per cent C_2H_5OH). An electron-acceptor group, e.g. NO_2 in the 4'-position,

† Note the fundamental difference as compared with phosphorescence (p. 6), where transition originates from a singlet and not from a triplet state.

produces a definite bathochromic shift ($\lambda = 478$ nm, $\varepsilon = 33{,}100$ l mol^{-1} cm^{-1} in 95 per cent C_2H_5OH). This analysis shows, however, that these intense visible bands do not correspond to the $n_s \rightarrow \pi^*$-transition but principally to the $\pi \rightarrow \pi^*$-transition of azobenzene (Table 2.1). On these are superimposed negligible $n_s \rightarrow \pi^*$ and, presumably, further transitions of lower extinction.

The colour rules developed by Witt[1] and extended by Wizinger[10] and others will now be briefly discussed in the light of the quantum chemical methods reviewed at the beginning of this Chapter. In accordance with these theories, a coloured compound is built up from three subsystems, namely an auxochrome and a chromophoric group linked together by a system of conjugated double bonds (e.g. aromatic hydrocarbons or polyenes). Both auxochromes and chromophores shift the higher wavelength absorption bands of the conjugated system to longer wavelengths (i.e. bathochromically) and both can be classified empirically in order of increasing bathochromic effect on a particular conjugated system. On the other hand, there are rules, relating bathochromic shift to position for both types of group mentioned above, for particular auxochromes and chromophores in a conjugated system.

As physical and organic chemistry has developed, particularly with regard to the investigation of the electronic effects of substituent groups, it has become apparent that auxochromes are *electron donors* and chromophores *electron acceptors*.

The rule that a conjugated compound exhibits intense colour only when it carries an auxochrome stems from the time when the common basis of ultraviolet and visible spectra was still unknown to chemists. The comparison of the spectrum of azobenzene with those of its 4-dimethylamino- and 4-dimethylamino-4'-nitro derivatives was mentioned above as it constitutes a good example.

Historically, the resonance theory is sandwiched between the rules postulated by Witt and the quantum mechanical electron theory of organic chemistry. Since the 1930's attempts have been made to treat the problem of constitution and absorption spectrum from this angle also. Thus it is possible to write an increasing number of polar limiting structures for azobenzene, Dimethyl Yellow and 4-dimethylamino-4'-nitroazobenzene, which are meant to symbolize the easy 'excitability'. Apart from the fact that the choice of the limiting structures is largely arbitrary, the recent treatment by quantum chemical approximation methods shows that the implicit assumption of equal distribution of limiting structures in the ground and excited states is incorrect.

Thus Grinter and Heilbronner[11] for instance were able to demonstrate, with the aid of the MIM method, for the spectra of the three *o*-, *m*- and *p*-isomers of nitrophenol, cyanophenol and nitro-N-dimethylaniline, that the charge distributions are not correctly represented by the mesomeric structures. Contrary to expectation from the theory of Witt, or from simple considerations of the mesomeric structures, the *o*- and *p*-isomers do not absorb at the longest wavelengths; the *o*- and *m*-compounds absorb very similarly over the whole domain above 220 nm, whereas the absorption bands of the *p*-isomers are not only hypsochromically displaced but also quite different in type. In contrast, the MIM methods lead in all cases to a good agreement between theory and experiment.

Wizinger[10] has also pointed out that it is impossible to understand the so-called distribution principle of auxochromes on the basis of the resonance theory. This refers to the fact that the introduction of a further donor into a benzene derivative that already carries one electron acceptor and a donor causes the following shifts relative to the 1,4-donor-acceptor compound:

(2.3)

Bathochromy

Here, too, it was seen[11] that the cause of this shift to long wavelengths lies in the fact that the energy in the excited state does not correspond to that which could be expected from the customary concepts of the resonance theory. The following formulae (2.4) give the summed charge displacements in percentages of a full electron charge for ground and excited states (GS, ES) of benzene derivatives (D = donor, A = acceptor) which were calculated by Grinter and Heilbronner.[11]†

The principles on which the distribution rules of the auxochromes are based are also of importance when extended to anthraquinone, disperse and vat dyes (sections 10.2 and 10.5), and for quinacridone (section 10.4) and other classes of dyes.

The old problem of explaining the strikingly dark blue colour of indigo in terms of its chemical structure (**2.1**) was solved by Klessinger and Lüttke[12] with the aid of the Pariser–Parr–Pople method (for indigo $R^1 = R^2 = R^3 = R^4 = H$). It is hardly conceivable that this knowledge could have been obtained experimentally since Lüttke and coworkers succeeded much later in synthesizing compounds which represent the parent structure responsible for the visible colour of indigo.

When reviewing the development of our knowledge of the connection between chemical constitution and light absorption since Witt postulated his colour rules in 1876, we realize that it is possible nowadays to elucidate quantitatively and not empirically the mechanisms which are responsible for the absorption of light. It is an indispensable consequence of the nature of the quantum theory that the acquirement of this knowledge involved a loss of simplicity and necessitated a considerable amount of work in respect of each specific problem.

† These numerical data refer to specific models where *definite* values for the MIM parameter were inserted. These values, of course, differ slightly in accordance with the type of donor and acceptor in specific compounds.

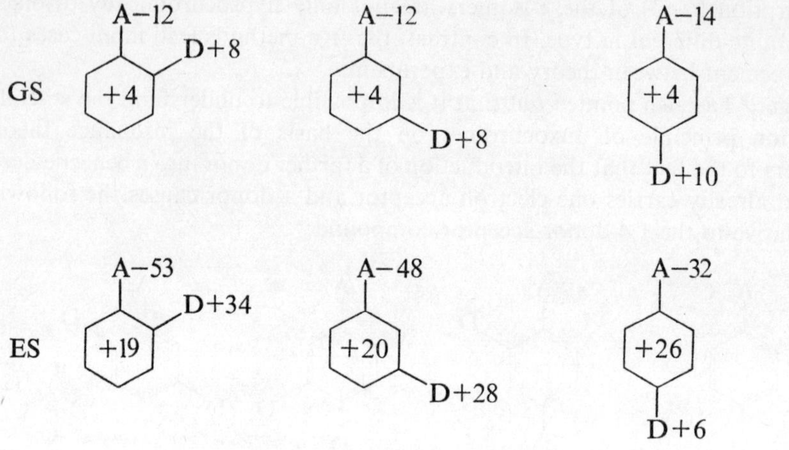

Fundamentals of the Chemistry and Application of Dyes

$$
\begin{array}{c}
\text{GS} \\
\text{ES} \\
\text{GS} \\
\text{ES}
\end{array}
$$

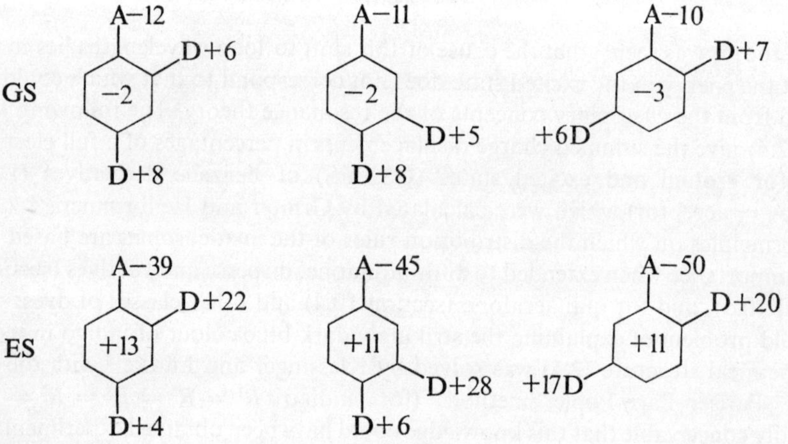

(2.4)

Structure (2.1):

$$
\begin{array}{c}
R^1-\underset{\parallel}{C}-\cdots-\underset{\mid}{N}-R^3 \\
O H \\
 C=C \\
R^2-\underset{\mid}{N}\underset{\parallel}{C}-R^4 \\
H O
\end{array}
$$

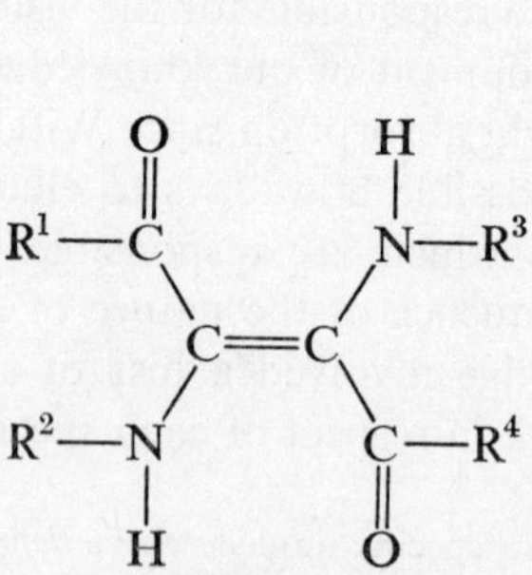

(2.1)

2.2. COLORIMETRY

An object appears coloured to the human eye either because it emits a portion of the visible spectrum (e.g. a sodium vapour lamp) or because light falling on it, e.g. sunlight, is partly absorbed and partly reflected. The reflected light is of complementary colour to the absorbed light, i.e. it is an additive mixture of all spectral colours with the exception of those absorbed.

Since, however, coloured objects never absorb completely any part of the spectrum, and absorption at each wavelength may vary from 0 to 100 per cent (corresponding to 100 to 0 per cent reflection), many mixed hues appear apart from the pure spectral colours: the trained eye of an artist or dyer can distinguish over 2 million. However, humans, in general, have poor colour memory and some means of describing colours quantitatively is therefore necessary.

Absorption or reflection spectra (plots of absorption or reflection as a function of wavelength or frequency over the visible spectrum) are unsuitable in this respect since the sensitivity of the human eye varies over the whole of the visible spectral region, with a pronounced maximum at 555 nm. Moreover, bodies with different absorption or reflection spectra can, under particular conditions of illumination, appear the same colour (so-called metameric colours).† Like all other sensations, the visual colour impression cannot be described in absolute terms but only by comparison: something is as sweet as sugar or as sweet as honey; a wasp sting hurts like a nettle sting, like a sprain; an object is as black as night, as a raven, as coal. This means that we characterize as the same, or similar, sensations produced by widely different stimuli.

The most widely adopted system of colour description is the CIE system (Commission International de l'Eclairage). This is based on the fact that light reflected from any coloured surface can be visually matched by a mixture of red, green and blue lights in suitable proportions. A system of such lights that will allow any colour to be matched *directly* is not, however, attainable in practice; to match spectral colours, for example, it is necessary first to add one of the real primaries to the colour to be matched, and then match the resulting colour with the remaining primaries.

The red, green and blue primaries of the CIE system are described by spectral response curves which were derived from the average results of colour matching experiments made by a number of observers having normal colour vision. This was necessary because the colour sensitivity of the eye varies from one individual to another. (There are, of course, people who are partially or completely 'colour blind'; these are excluded from the present discussion.) The response curves most commonly used refer to a matching field subtending an angle of 2° at the eye and

---

† Metamerism plays a part in the application of dyes and the problem of the so-called evening colours: metameric dyeings which are visually identical in colour in daylight show distinct colour differences when illuminated with a light source of different spectral composition (e.g. a filament lamp).

define the 1931 CIE standard observer. More recently data for a 10° visual field have become available (1964 CIE 10° standard observer). The difference between the two sets of results arises from the fact that the 2° field involves the retinal cones only whereas the 10° field involves both rods and cones. Rods and cones differ greatly in their sensitivity to colour.

The CIE standard observer response curves $\overline{X}$(red), $\overline{Y}$(green) and $\overline{Z}$(blue) give the amounts, $\overline{X}_\lambda$, $\overline{Y}_\lambda$, $\overline{Z}_\lambda$, of the red, green and blue primaries, respectively, required by the Standard Observer to match light of wavelength λ. In addition, $\overline{Y}$ is arranged to correspond exactly to the average luminous sensitivity curve for an average eye. Thus, the CIE description of a colour can be derived from its reflectance spectrum by multiplying the percentage reflectance at each wavelength R_λ by the appropriate values of $\overline{X}_\lambda$, $\overline{Y}_\lambda$, $\overline{Z}_\lambda$. It is also necessary to take account of the energy distribution of the light in which the colour is to be viewed, viz. $\bar{E}_\lambda$. The CIE tristimulus values are then:

$$X = \sum_{400}^{700} E_\lambda \cdot R_\lambda \cdot \overline{X}_\lambda$$

$$Y = \sum_{400}^{700} E_\lambda \cdot R_\lambda \cdot \overline{Y}_\lambda$$

$$Z = \sum_{400}^{700} E_\lambda \cdot R_\lambda \cdot \overline{Z}_\lambda$$

Y is also a direct measure of luminosity (luminous reflectance). For a perfect black $Y = 0$ and for a perfect white $Y = 100$.

Photoelectric colorimeters are available that give X, Y and Z directly.

Graphical presentation of tristimulus values would require a three-dimensional coordinate system, and so three new quantities, the chromaticity co-ordinates, are derived from X, Y and Z as follows:

$$x = \frac{X}{X + Y + Z} \qquad y = \frac{Y}{X + Y + Z} \qquad z = \frac{Z}{X + Y + Z}$$

Since $x + y + z = 1$, x and y alone are sufficient to describe the colour, disregarding luminosity.

A diagram with x as abscissa and y as ordinate is termed the CIE chromaticity diagram (Figure 2.4). The spectral colours lie on an almost parabolic curve, the ends of which are connected by a straight line representing purples. Introduction of the luminosity coordinate perpendicular to the diagram leads to the CIE colour solid. The neutral colours (white, grey, black) lie on a line perpendicular to the base, near the centre of the solid.†

By means of the two-dimensional CIE diagram, two other characteristics of a

---

† The position of the neutral point varies in accordance with the spectral composition of the illuminant (daylight, filament lamp, etc.).

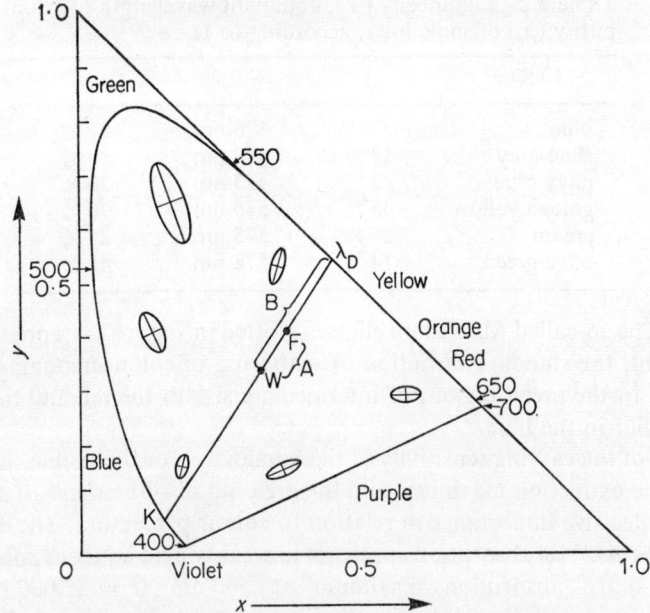

Figure 2.4 CIE chromaticity diagram according to Hardy.[13] W = neutral point (white), F = colour point of the dominant wavelength λ_D, K = complementary colour. MacAdam ellipses: compare text with page 18.

colour may be determined, namely dominant wavelength (λ_D) and purity (p_e) (Figure 2.4). The dominant wavelength is given by the point at which a straight line drawn from the neutral point (W) through the colour point (F) cuts the spectral colour curve. The purity is the greater the nearer the colour point lies to the latter. Extrapolation of the straight line in the other direction gives the dominant wavelength of the complementary colour (K). In luminosity, dominant wavelength and purity we have three physically-measurable quantities that correspond roughly to the subjective properties commonly used to describe a colour, viz. brightness, hue and saturation.

The examples of Table 2.2 show, for two groups of colours, that the visual colour impression is influenced by luminosity and purity despite the approximately equal dominant wavelengths.

In contrast with previous attempts to devise a colour specification, the CIE system takes into account the differing sensitivity of the eye to different parts of the spectrum. That the system is, however, based on the measurement of stimuli and not perception, is shown by the fact that equal visual differences between pairs of colours in regions of different hue do not correspond to equal distances in the CIE

Table 2.2 Luminosity (Y), dominant wavelength (λ_D) and purity (p_e) of some hues, according to Hardy[13]

Colour	Y	λ_D	p_e
blue	32%	476 nm	27%
slate grey	12%	476 nm	10%
navy blue	3%	475 nm	20%
golden yellow	45%	576 nm	70%
cream	55%	575 nm	25%
olive green	14%	572 nm	45%

diagram. The so-called MacAdam ellipses plotted in Figure 2.4 represent, in 10-fold enlargement, the standard deviation of scattering of colour matching. The ellipses are largest in the green region, of intermediate size in the red and the yellow, and much smaller in the blue.

Because of the varying sensitivity of the human eye for the visible spectral region, not only the extinction maximum and the area but also the *shape* of an absorption band is of decisive importance in relation to colour perception. The dye (2.2) is an example of this.[15] Its absorption spectrum in weakly acid aqueous solution shows a relatively sharp absorption maximum at 500 nm ($\varepsilon = 25,000 \, l \, mol^{-1} \, cm^{-1}$). The hydroxyl group dissociates in alkaline solution. The naphtholate ion has a broad absorption band with a maximum at 478 nm ($\varepsilon = 14,000 \, l \, mol^{-1} \, cm^{-1}$). This band extends to 700 nm.

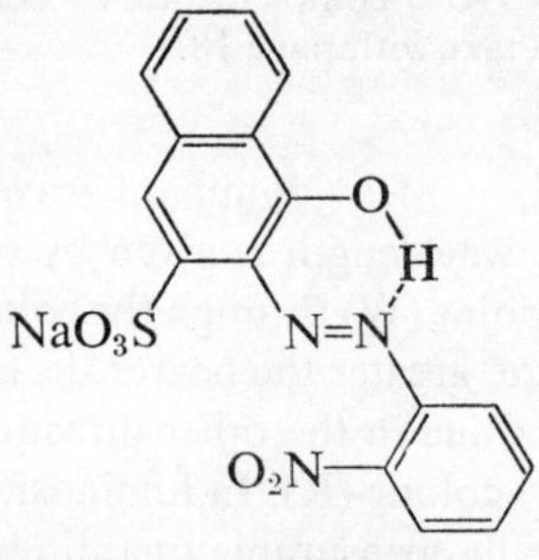

(2.2)

From the absorption maximum alone, an orange-red colour in acid solution and a yellowish-orange hue in alkaline solution would be expected. However, a colour change from orange to dull wine red is observed! Investigations regarding the relation between chemical constitution and light absorption should therefore always be based on measured spectra and not visual colour assessment.

In this example the visual impression does not correspond to the hypsochromic shift of the maximum because there is a significantly larger absorption by the naphtholate ion in the region of maximum visual sensitivity (555 nm) than with the naphthol which has a sharp peak in the long wavelength region.

The parameters of the CIE system express this very neatly (Table 2.3): whilst the dominant wavelengths are almost the same in both cases the naphtholate is 20 per cent darker than the naphthol as shown by the Y value, and its dull shade is shown by the lower value for p_e.

Table 2.3 CIE data for dye (2.2)[15] [a]

	λ_{max}	Colour expected[b]	λ_D	Y	p_e	Colour observed
Naphthol	500 nm	orange red	591 nm	67%	43%	orange
Naphtholate	478 nm	orange yellow	592 nm	54%	32%	dull wine red

[a] Concentration 4×10^{-5} mol/l, 10 mm thickness of layer.
[b] The colour to be expected on the basis of λ_{max} only.

Although the CIE system can give an unequivocal description of colours it does not suffice to define tolerances among specimens differing in hue. This problem can be solved only with a system based on the measurement of sensations.[14,16]

According to the Weber–Fechner law there is a relation between sensations and the physical stimuli on which they are based. The difference between two sensations ΔE is proportional to the difference ΔR in the stimulus relative to the magnitude of the latter, i.e. $\Delta R/R$. On the other hand the MacAdam colour difference ellipses (or ellipsoids in the chromaticity space) vary with hue in size and orientation so that it is not possible directly to derive a system based on the measurement of sensitivity. The CIE is attempting to select a visually uniform chromaticity diagram. Replacement of the chromaticity coordinates x and y with the coordinates $u = 4X/(X + 15Y + 3Z)$ and $v = 6Y/(X + 15Y + 3Z)$ gives a fair approximation when colours of approximately equal luminosity Y are to be compared (CIE 1960 UCS diagram, UCS = uniform colour space).

The most rigorous investigations that have been carried out concerning the problem of uniform colour spacing were based on the colour system of the American painter Munsell, which was later modified by various workers.

The Munsell solid is represented by a series of charts, each of which corresponds to a particular hue, e.g. yellow, red, yellow-red. The charts are arranged radially within a cylinder, with equiangular spacing (visually equal steps), all the inner edges lying on the axis of the cylinder which represents neutral colours ranging from black through grey to white. Upon each chart, coloured chips are arranged such that all in the same horizontal row are of equal lightness (Munsell term: value) but vary in degree of saturation (Munsell term: chroma) in visually equal steps with the least saturated colours at the inner edge. Corresponding rows of samples on different charts also have the same lightness (value). The chips are also so arranged that those equidistant from the axis, i.e. in corresponding vertical columns, have the same chroma but vary in lightness from very dark (black in the neutral column at the bottom of the axis) to very light (white at the top of the axis). The number of chips in the columns of course decreases with increase in saturation, and the number in the rows varies with lightness, the position of the row containing most chips (i.e. including the chip of highest chroma) varying with colour, being at a high value for a 'light' colour such as yellow and at a low value for a 'dark' colour such as blue.

A relation can be derived between the Munsell value (V) and the CIE tristimulus value Y. Adams[17] and Nickerson[17] have extended this treatment to X and Z. The resulting functions, V_X, V_Y, V_Z, have been used to define a visually uniform colour space in which total colour differences Δe (i.e. differences between samples differing in lightness as well as hue) are calculated by means of the equation

$$(\Delta e)^2 = c^2\{(0.23\,\Delta V_Y)^2 + [\Delta(V_X - V_Y)]^2 + [0.4\,\Delta(V_Z - V_Y)]^2\}$$

A given value of Δe is claimed to have the same visual significance in any part of colour space. The constant c is included to make the unit of colour difference equal *on average* to the NBS unit. The best value for c has been given as 41·86 but for practical purposes an approximate value, e.g. 50, may be used. Differences of 1 NBS unit or less are often considered unimportant in commercial colour matching but in the dyeing industry even smaller differences may be unacceptable. In certain regions of colour space differences very much smaller than 1 NBS unit, perhaps as small as 0·2 NBS unit, can easily be seen at least by a trained colourist.

Other attempts to devise uniform colour space systems, e.g. the CIE 1964 UCS system, have been described in the standard work by Wyszecki and Stiles.[14]

The problem of instrumental recipe prediction is even more complicated as the relation between the reflectance of a dyeing and dye concentration is not linear.

The Kubelka and Munk theory[18] leads to the equation

$$K/S = \frac{(1-R)^2}{2R} \tag{2.5}$$

where K is the coefficient of absorption, S the coefficient of scatter and R the reflectance of the sample at a given wavelength. To a first approximation K and S are proportional to the concentration of each dye in the substrate and if each dye in a mixture acts additively (this is not always the case) we have

$$K/S_M = \frac{aK_A + bK_B + cK_C + K_W}{aS_A + bS_B + cS_C + S_W} \tag{2.6}$$

where the subscript M refers to the mixture, A, B and C refer to the different dyes, and a, b, c represent the concentration of each dye. K_W and S_W are the absorption and scattering coefficients of the substrate. Since S_W is normally large compared with the scattering caused by the dyes the above equation reduces to

$$K/S_M = \frac{aK_A}{S_W} + \frac{bK_B}{S_W} \ldots + \frac{K_W}{S_W} \tag{2.7}$$

This equation has been used as the basis of a system of recipe prediction in which values of K/S at different wavelengths are used to provide sets of simultaneous equations that can be solved for dyestuff concentrations by an analogue computer; other systems employing different equations and digital computers have also been developed.[19]

A description of the physiology of colour vision of animal and man is not within the scope of this textbook. In addition to the highly sensitive rod cells which are responsible for 'colourless' night vision, the retina of human beings contains three types of cone cells which exhibit sensitivity maxima for different spectral regions (440, 540 and 567 nm). With bees, however, the sensitivity range lies at lower wavelengths; they can see ultraviolet 'colours' down to 300 nm (K. v. Frisch). In certain respects these three types of receptors correspond to the physical colour theory on which the CIE system, described above, has been based. This three-component theory of colour vision was postulated by T. Young in 1802 and further developed by H. Helmholtz (1852) and J. C. Maxwell (1857). E. Hering in 1875 postulated the so-called opponent colour theory according to which colour vision is based on paired stimuli, i.e. white-black, red-green and blue-yellow. Physiological investigations in the 20th century appear to justify the Young–Helmholtz–Maxwell theory. Only recently, neuro-physiological experiments, in particular with the eyes of fish (which also contain three types of cones), have shown that light signals absorbed by the retina in accordance with a three-component mechanism are processed in the neurons in accordance with an opposing component mechanism. From a cybernetic point of view this method must be based on extremely complicated processes.[20] The colour vision of animals and human beings is particularly remarkable because the eye cones are sensitive to relatively broad spectral regions and therefore, from a physical point of view, cannot work precisely. Nevertheless, the eye gives a very accurate impression of our surroundings.

2.3. LITERATURE

1. O. N. Witt, *Ber. dtsch. chem. Ges.*, **9**, 522 (1876), **21**, 325 (1888).
2. A. Maccoll, *Quart. Rev. (chem. Soc. London)*, **1**, 16 (1947).
3. E. Hückel, *Z. Physik*, **70**, 204 (1931), **76**, 628 (1932); *Z. Elektrochem.*, **61**, 866 (1957); S. F. Mason, *J. Soc. Dyers Colourists*, **84**, 604 (1968); E. Heilbronner and H. Bock, *Das HMO-Modell und seine Anwendung*, Vol. 3, Verlag Chemie, Weinheim, 1968–70; G. M. Badger, *Aromatic Character and Aromaticity*, University Press, Cambridge, 1969; M. J. S. Dewar, *The Molecular Orbital Theory of Organic Chemistry*, McGraw-Hill, New York, 1969.
4. Summary: H. Kuhn, *Experientia (Basel)*, **9**, 41 (1953); N. S. Bayliss, *Quart. Rev. (chem. Soc. London)*, **6**, 319 (1952); J. R. Platt and coworkers, *Free-Electron Theory of Conjugated Molecules*, University of Chicago Press, 1964.
5. H. Kuhn, *Helv. chim. Acta*, **31**, 1441 (1948), **32**, 2247 (1949), **34**, 2371 (1951); J. R. Platt, *J. chem. Physics*, **17**, 484, 1198 (1949).
6. J. E. Lennard-Jones, *Am. Rev. physic. Chem.*, **4**, 167 (1953).
7. R. Pariser and R. G. Parr, *J. chem. Physics*, **21**, 466, 767 (1953); J. A. Pople, *Trans. Faraday Soc.*, **49**, 1375 (1959).

8. H. C. Longuet-Higgins and J. N. Murrell, *Proc. physic. Soc.* **A68**, 601 (1955); W. T. Simpson, *Theories of Electrons in Molecules*, Prentice-Hall, Englewood Cliffs, 1962.
9. G. Kortüm and H. Rau, *Ber. Bunsenges. physik. Chem.*, **68**, 973 (1964); W. Maier, A. Saupe and A. Englert, *Z. physik. Chem. (Frankfurt a.M.)* **10**, 273 (1957); H. H. Jaffé, S. J. Yeh and R. W. Gardner, *J. molecular Spectroscopy*, **2**, 120 (1958); H. Bock, *Angew. Chem.*, **77**, 469 (1965).
10. R. Wizinger, *Organische Farbstoffe*, Bonn, 1933; *Chimia (Aarau, Schweiz)*, **15**, 89 (1961), also further references given therein.
11. R. Grinter and E. Heilbronner, *Helv. chim. Acta*, **45**, 2496 (1962); H. Labhart and G. Wagnière, *Helv. chim. Acta*, **46**, 1314 (1963).
12. M. Klessinger and W. Lüttke, *Tetrahedron (London)*, **19**, Suppl. 2, 315 (1963); M. Klessinger, *Tetrahedron (London)*, **22**, 3355 (1966).
13. Summary: A. C. Hardy, *Handbook of Colorimetry*, M.I.T. Press, Cambridge (Mass.) 1936; W. Schultze, *Farbenlehre und Farbenmessungen*, 2nd edition, Springer-Verlag, Berlin, 1966; O. Hayer, *Einfärben von Kunststoffen*, Hanser-Verlag, München, 1962.
14. G. Wyszecki and W. S. Stiles, *Color Sciences*, John Wiley, New York, 1967.
15. O. A. Stamm and H. Zollinger, *Verh. Naturf. Ges. Basel*, **67**, 367 (1956); H. Zollinger, *Chemie der Azofarbstoffe*, Birkhäuser-Verlag, Basel, 1958, section 12.2
16. Summary: E. Ganz, *Textil-Rdsch.*, **18**, 242 (1963); U. Gugerli, *Textil-Rdsch.*, **18**, 252 (1963).
17. E. Q. Adams, *J. Opt. Soc. Am.*, **32**, 168 (1942); D. J. Nickerson, *J. Opt. Soc. Am.*, **34**, 550 (1944), *Am. Dyest. Rep.*, **39**, 541 (1950).
18. P. Kubelka and F. Munk, *Z. techn. Phys.*, **12**, 593 (1931).
19. J. V. Alderson, E. Atherton and coworkers, *J. Soc. Dyers Col.*, **77**, 657 (1961), **79**, 723 (1963); U. Gugerli and coworkers, *J. Soc. Dyers Col.*, **79**, 637 (1963), *Textilveredlung*, **1**, 29 (1966); E. Ganz, *Textil-Rdsch.* **20**, 255 (1965). D. Strocka in *Ullmann's Encyclopädie der technischen Chemie*, Urban and Schwarzenberg, München, 1970, Supplementary Volume, 233.
20. Summary: H. Autrum, *Naturwissenschaften*, **55**, 10 (1968). See also C. H. Graham, *Vision and Visual Perception*, John Wiley, New York, 1965, also B. Hassenstein, *Kybernetik*, **4**, 209 (1968); K. Motokawa; *Physiology of Color and Pattern Vision*, Igaku Shoin, Tokyo, and Springer, Berlin, 1970; T. N. Cornsweet, *Visual Perception*, Academic Press, London, 1971.

3

The Fundamentals of the Reaction Mechanisms of Dyestuff Syntheses

3.1. INTRODUCTION

As section 2.1 indicates, dyes absorb visible light for the same reasons that they, and many other compounds, absorb ultraviolet light. From a purely chemical point of view it would therefore not be sensible to call colour chemistry a separate section of pure chemistry. Colour chemistry is, however, a branch of applied chemistry since it deals with the synthesis of coloured compounds which can be used for dyeing various substrates. From the point of view of applied chemistry *synthesis* is therefore the most important part of colour chemistry. In industry synthetic methods that are as efficient as possible with respect to yield, purity of product and economy are essential.

The scientific basis of a modern, non-empirical optimization of a synthesis is an understanding of the mechanism of each reaction step. The elucidation of a reaction mechanism permits us to recognise which types of equilibria of the reactants interact with one another, the role of the solvent, whether the reaction can be catalysed and, if so, the structural features of an optimal catalyst.

These investigations are the object of the so-called physical-organic chemistry. In addition to the physical methods of structural analysis (spectra of all types), kinetic methods play a large part.

The kinetic results from such work are directly useful in optimizing processes as well as yielding general information about the reaction mechanism. An increase of the reaction velocity is of interest technically, not only because it can, in certain circumstances, make more preparations possible in a given time, but, mainly, because it can lead to improved yields and purity. For instance, it is easy to demonstrate kinetically that in a reaction which yields isomers that consume 50 per cent of the reagent, as well as the desired main product, the yield of the latter can be raised from 50 to 80 per cent if the rate of the main reaction can be increased by a factor of four or that leading to the formation of the isomers decreased correspondingly, perhaps by using suitable catalysts.

A comprehensive modern scientific treatment of colour chemistry, therefore, requires a sound knowledge of physical-organic chemistry. There is not sufficient space to give a thorough presentation within the framework of this manual. There is, however, a series of good monographs available in this sphere of physical-organic chemistry.[1] In the present book the authors will merely outline those domains that are of particular importance in colour chemistry, namely aromatic substitution (sections 3.2 to 3.4), the formation of metal complexes (section 3.6) and the redox equilibria and reactions in which quinones participate (section 3.5), together with the essential features of the application of physical-organic chemistry to technical problems. In subsequent chapters, where separate classes of dyes are treated from the structural point of view, it is then possible to refer to these fundamentals.

It should be pointed out at this point that 'intermediate' in the technological sense means a product (with preparatory and technical importance, and usually isolated) of an intermediate step in a multistage synthesis (cf. ref. 1a), whereas the physical-organic term 'intermediate' refers to a thermodynamically stable particle (in contrast to a transition state, cf. section 3.3) postulated in a reaction mechanism, irrespective of its being isolated or detected in a measurable concentration or being present only as a steady-state intermediate.

3.2. PREEQUILIBRIA IN AROMATIC SUBSTITUTION

The classification of aromatic substitution (Table 3.1) reactions is based essentially on the system devised by C. K. Ingold.[2] With heterolytic substitutions, both electrons of the bond between the aromatic substrate and the substituent, which is introduced, originate either from the substrate or from the reagent (electrophilic or nucleophilic substitution respectively). In a homolytic or radical substitution one electron each is provided by the reagent and by the substrate.

Table 3.1 Classification of the substitution reactions

Type of bond formation	Heterolytic substitution					Homolytic substitution
Type of reagent	Electrophilic		Nucleophilic			Radical
Mechanistic type	S_E2	σ-complex	S_N2-type	Addition–elimination mechanism	Elimination–addition mechanism (arin)	Elimination–addition mechanism of radicals
Example	Nitration	Amination with NCl_3	Formation of phenols from diazonium ions	Aminolysis	Dow phenol process	Side-chain chlorination of toluene

Electrophilic aromatic substitutions in accordance with the S_E2 mechanism, which have been rigorously investigated preparatively, technically and mechanistically, provide the most important route to derivatives of aromatic hydrocarbons. Reactions

The Fundamentals of the Reaction Mechanisms of Dyestuff Syntheses

based on the so-called σ-mechanism, which was only discovered in 1964, have so far not achieved any practical significance.[3]

With S_E2-substitutions the equilibria which precede the so-called substitution proper play a large part, since the electrophilic reactant and the nucleophilic substrate are not present under all reaction conditions (solvent, pH etc.) in a form in which both partners can react with one another in the substitution proper. Both belong to a system of acid–base equilibria; usually only one, and never all, of the separate equilibrium stages is effective in the substitution proper.

It is a general rule that the most reactive particle of the electrophilic component is the equilibrium form with the smallest pK_a value and that conversely the nucleophilic substrate has the greatest reactivity in the most strongly basic equilibrium form.

This indicates that solvents, catalysts, additives, buffers, etc., must, if possible, be so chosen that the most reactive particles are present in optimum concentrations in *both* equilibrium systems.

The elucidation in 1950 by Ingold, Hughes and collaborators of the mechanism of nitration is a classical example of the investigation of the most reactive types of equilibria in an electrophilic aromatic substitution. It is seen from Table 3.2 that the nitryl ion, a Lewis acid, is the most reactive nitrating reagent. As a general rule that holds for all electrophilic reagents, all electrophilic particles can be derived from the corresponding Lewis acid $X^\oplus$ in which the Lewis acid is covalently bound to bases Y or $Y^\ominus$: $X^\oplus + Y^\ominus \rightarrow X\text{—}Y$, or $X^\oplus + Y \rightarrow X\text{—}Y^\oplus \leftrightarrow X^\oplus\text{—}Y$. Here the reactivity of the electrophilic reagent falls with increasing basicity of Y or $Y^\ominus$. Which species acts as electrophilic reagent in a particular case depends not only on the sequence of the reactions but also on the equilibrium concentrations. Therefore, it is necessary to determine the reacting species experimentally for each separate case.

Electrophilic reagents can replace not only aromatic hydrogen atoms but also those that are combined with nitrogen and oxygen. The replacement of hydrogen

Table 3.2 Acid–base equilibria of nitric acid

$$\overset{\oplus}{N}O_2 \underset{\longleftarrow}{\overset{+H_2O}{\longrightarrow}} H_2\overset{\oplus}{O}\text{–}NO_2 \underset{\longleftarrow}{\overset{-H^\oplus}{\longrightarrow}} HO\text{–}NO_2 \underset{\longleftarrow}{\overset{-H^\oplus}{\longrightarrow}} \overset{\ominus}{O}\text{–}NO_2$$

Name:	Nitryl ion	Nitroacidium ion	Nitric acid molecule	Nitrate ion
Type:	Lewis acid	Brönsted acid or conjugate acid of the nitric acid molecule	Brönsted acid or conjugate base of the nitroacidium ion	Base
Reactivity:		$\overset{\oplus}{N}O_2 > H_2\overset{\oplus}{O}\text{–}NO_2 > HO\text{–}NO_2 > \overset{\ominus}{O}\text{–}NO_2 = 0$		
Acidity:		$\overset{\oplus}{N}O_2 > H_2\overset{\oplus}{O}\text{–}NO_2 > HO\text{–}NO_2 > \overset{\ominus}{O}\text{–}NO_2$		

and primary aromatic amino groups by the nitroso group (diazotization, cf. section 4.1.2) is of special importance in dyestuff chemistry. The diazonium ion thus arising acts as electrophilic component in azo coupling (cf. section 4.1.3). This is the most important reaction in the production of azo dyes (cf. below and Chapter 4).

The preequilibria of the electrophilic components in nitrosations are equivalent to those in nitration with the difference that the nitrosyl ion ($NO^{\oplus}$) corresponding to the nitryl ion acts as nitrosation reagent only in extreme conditions. Equations (3.1) and (3.2) show that the nitrosoacidium ion $H_2\overset{\oplus}{O}$—NO which corresponds to the nitroacidium ion in nitration does not normally split off water with formation of the nitrosyl ion but it forms, with a base $Y^{\ominus}$ of the system, the addition compound Y—NO which subsequently attacks (as electrophilic reagent) the aromatic amine (as nucleophilic substrate) (cf. section 4.1.2).[4]

$$HNO_2 + H^{\oplus} = H_2\overset{\oplus}{O}\text{—NO} \tag{3.1}$$

$$H_2\overset{\oplus}{O}\text{—NO} + Y^{\ominus} = Y\text{—NO} + H_2O \tag{3.2}$$

$$Y^{\ominus} = NO_2^{\ominus} \text{ (in dilute } HClO_4 \text{ and in } <85\% \ H_2SO_4\text{)}$$
$$= Cl^{\ominus}, Br^{\ominus} \text{ (in dilute HCl or HBr)}$$
$$= OSO_3H^{\ominus} \text{ (in } >85\% \ H_2SO_4\text{)}$$

Since nitrosation reagents are much weaker particles than the corresponding nitrating agents, with nitrosations greater attention must be paid to the position of the preequilibria, not only for the electrophilic component, but also for the nucleophilic substrate: the aromatic amine participates only as free base (Ar—NH_2), not as ammonium ion (Ar—$\overset{\oplus}{N}H_3$) in the substitution proper (cf. section 3.3).

$$Ar\text{—}NH_2 + H^{\oplus} \leftrightarrows Ar\text{—}\overset{\oplus}{N}H_3 \tag{3.3}$$

The diazonium ion is a Lewis acid; it is converted into diazohydroxide, i.e. a Brönsted acid, by the addition of a hydroxyl ion; the diazotate results from the splitting of a proton from the diazohydroxide. References 4–8 show that the diazonium ion must therefore be regarded as a dibasic acid.

1st Step

$$[\text{Ar}\text{—}\overset{\oplus}{N}\equiv\ddot{N} \leftrightarrow \text{Ar}\text{—}\ddot{N}=\overset{\oplus}{N}] + OH^{\ominus} \leftrightarrows ArN_2OH \tag{3.4}$$
$$\text{Lewis acid} \qquad\qquad\qquad \text{Diazohydroxide}$$

or
$$ArN_2^{\oplus} + H_2O \underset{}{\overset{K_1'}{\rightleftarrows}} ArN_2OH + H^{\oplus} \text{ respectively} \tag{3.5}$$

$$K_1 = K_1'[H_2O] = \frac{[ArN_2OH][H^{\oplus}]}{[ArN_2^{\oplus}]} \tag{3.6}$$

2nd Step

$$\text{ArN}_2\text{OH} \underset{}{\overset{K_2}{\rightleftarrows}} \text{ArN}_2\text{O}^\ominus + \text{H}^\oplus \qquad (3.7)$$
$$\text{Diazotate}$$

$$K_2 = \frac{[\text{ArN}_2\text{O}^\ominus][\text{H}^\oplus]}{[\text{ArN}_2\text{OH}]} \qquad (3.8)$$

With dibasic Brönsted acids (e.g. H_2SO_4, oxalic acid, resorcinol) the first acidity constant (K_1) is considerably larger than the second constant (K_2). Exceptions to this rule are possible when the dibasic acid is a Lewis acid, but the monobasic acid is a Brönsted acid. In these cases K_2 can be larger than K_1. This is so for the diazonium ion. It has been demonstrated that for most diazonium ions investigated K_2 is more than 1,000 times as large, in aqueous solution, as K_1.[5] Hence these two constants cannot be determined separately, potentiometrically or otherwise, as when $K_1 > K_2$, and only the quantities $(K_1 K_2)^{1/2}$ or $(pK_1 + pK_2)/2$ can be measured.

It follows from the relation $K_2 > K_1$ between the two acidity constants that the monobasic acid (diazohydroxide) cannot be the most stable species in any pH range. The neutralization of one equivalent of a diazonium salt in aqueous solution with one equivalent of caustic soda does not, therefore, give rise to one equivalent of diazohydroxide but to one half-equivalent of diazotate and one half-equivalent of diazonium ion. Consequently the equilibrium concentration of the diazonium ion decreases above a pH value numerically equal to the quantity $(pK_1 + pK_2)/2$ not, as with the monobasic ion, tenfold but one hundredfold per pH unit.

Reversible *cis–trans*-rearrangements are, moreover, superimposed on the acid-base-equilibria of the diazo compounds. When an aqueous solution is heated a diazotate rearranges into a more stable isomer that no longer exhibits the typical reactions of diazo compounds. It is possible to regenerate the original reactivity of the diazo compounds by the addition of acids. Analogous reversible rearrangements were subsequently also found in respect of diazocyanides, diazosulphonates and azobenzene derivatives. The more stable isomer can often be converted photochemically into the less stable one.

There was a dispute lasting several decades between E. Bamberger, A. Hantzsch and others on the subject of the structure of these isomeric pairs of compounds. Hantzsch[6] postulated that the diazotates were pairs of geometrical (*cis/trans*) isomers which he termed *syn-* and *anti-*diazotates. This hypothesis was confirmed by J. M. Robertson by means of X-ray structural analyses of *cis-* and *trans-*azobenzene, by R. J. W. Le Fèvre who determined the dipole moments of isomeric diazocyanides and by Lüttke[7] for the particularly important case of diazotate isomers, who interpreted the infrared spectra of ^{15}N labelled isomers. Although there is still no direct evidence for the existence of diazosulphonate isomers it is to be assumed that these are also geometrical isomers. The terminology *syn/anti* as well as Bamberger's terms (normal or iso respectively) can therefore be replaced, nowadays, by *cis* and *trans* respectively.

With diazoamino compounds only *trans*-isomers are known so far.

As already mentioned, the diazonium ions formed during diazotization give rise to azo compounds in a subsequent electrophilic aromatic substitution of a nucleophilic substrate ('coupling component') in so-called *azo coupling* (section 4.1.3.). As diazonium ions are relatively weak electrophilic reagents, only aromatic compounds that carry electron donor substituents (OH, NH_2, NHR, etc.)[8] can normally be used as coupling components.

In carrying out azo couplings it is therefore necessary to take into account, not only these preequilibria of the diazo compounds, but also those of the coupling components. In accordance with the above-mentioned rule, i.e. that the reactivity of the nucleophilic substrate increases with rising basicity, the phenolate ion (Ar—$O^\ominus$) and the free amine (Ar—NH_2) must react more readily as coupling components than free phenol (Ar—OH) and the ammonium ion (Ar—$NH_3^\oplus$). This can be demonstrated experimentally by the dependence of the coupling kinetics on the hydrogen ion concentration of the medium.[9]

$$\text{Ar—OH} \rightleftarrows \text{Ar—O}^\ominus + \text{H}^\oplus \quad pK_a \text{ approximately 10} \qquad (3.9)$$

$$\text{Ar—NH}_3^\oplus \rightleftarrows \text{Ar—NH}_2 + \text{H}^\oplus \quad pK_a \text{ approximately 4} \qquad (3.10)$$

As an example, the rate of coupling of 4-toluenediazonium ion with 2,6-naphtholsulphonic acid is plotted against the pH value of the reaction medium in Figure 3.1.[8] Below pH 9 the coupling rate increases in direct proportion to the hydroxyl ion concentration since the equilibrium (3.9) of the coupling components (pK_a = 8.94) here shifts in the direction of the naphtholate ion (Ar—$O^\ominus$). Above pH 13

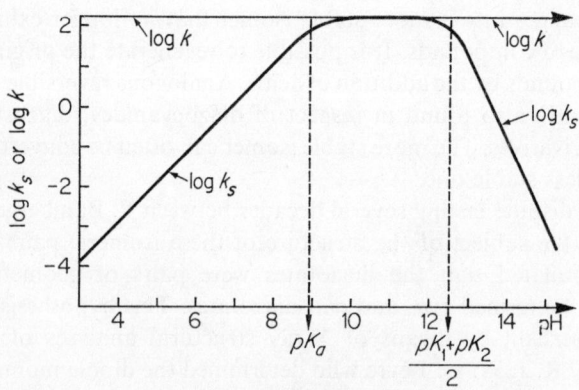

Figure 3.1 Rate of azo coupling of 4-toluenediazonium ion with 2,6-naphtholsulphonic acid as a function of the hydrogen ion concentration.[8]

The Fundamentals of the Reaction Mechanisms of Dyestuff Syntheses 29

the rate of coupling diminishes in inverse proportion to the square of the hydroxyl concentration because in this acidity range of the two-stage equilibrium system (3.4)–(3.8) the diazonium ion is transformed into the *cis*-diazotate. A rearrangement into *trans*-diazotate cannot be considered with this diazo compound at the reaction temperature of 0°C.

Apart from aromatic amines and phenols, enols also play a part in azo coupling as well as in other electrophilic substitutions in colour chemistry (e.g. colour development and colour photography, cf. sections 6.5 and 8.2). These include, for example, aceto acetanilides and 3-methyl-1-phenyl-5-pyrazolone derivatives. The question, which was fervently discussed in former years, whether these compounds react as enols or ketones is now superfluous since it is known that only the mutual conjugate base of enol *and* ketone of the equilibrium (3.11) participates in the substitution step proper.

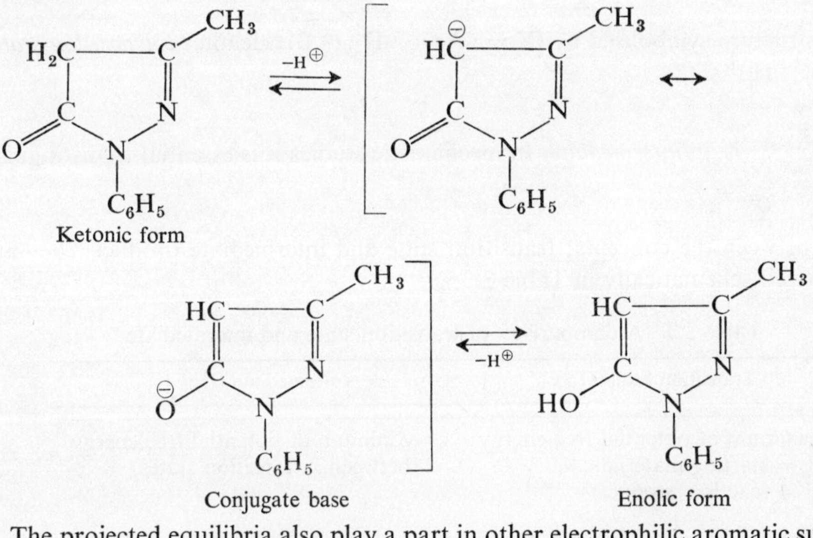

Ketonic form

Conjugate base   Enolic form   (3.11)

The projected equilibria also play a part in other electrophilic aromatic substitutions which are discussed in this book (e.g. the synthesis of triphenylmethane and methine dyes). However, they have so far been investigated only sporadically.

3.3. MECHANISM OF S_E2 REACTIONS

The investigation of the mechanism of the so-called substitution proper involves the question of how the reactive species of the electrophilic reagent and the nucleophilic substrate interact with one another. Two possible methods of substitution of a proton by an electrophilic reagent must be discussed:

(a) S_E3 mechanism: the addition of the electrophilic reagent and the transfer of the hydrogen as proton to a proton acceptor (Base B) occurs in a single step.

$$X^\oplus + Ar-H + B \rightarrow (X \cdots Ar \cdots H \cdots B)^\oplus \rightarrow Ar-X + H-B^\oplus$$

(3.12)

(b) S_E2 mechanism: in a first step (3.13) the electrophilic reagent is attached to form an intermediate product. In a subsequent step (3.14) the proton is transferred to the proton acceptor.

$$X^{\oplus} + Ar-H \underset{k_{-1}}{\overset{k_1}{\rightleftarrows}} Ar\overset{H}{\underset{X}{\diagdown}}^{\oplus} \qquad (3.13)$$

$$Ar\overset{H}{\underset{X}{\diagdown}}^{\oplus} + B \xrightarrow{k_2} Ar-X + H-B^{\oplus} \qquad (3.14)$$

The structure symbolised by $(X \cdots Ar \cdots H \cdots B)^{\oplus}$ denotes a *transition state*, while $Ar\overset{H}{\underset{X}{\diagdown}}^{\oplus}$ is an *intermediate*. In mechanistic studies it is essential to distinguish clearly between the concepts: transition state and intermediate product; they are summarized schematically in Table 3.3.

Table 3.3 A comparison of transition state and intermediate

Transition State (TS)	Intermediate (I)
Maximum of potential free energy (R = starting materials; P = reaction product)	Minimum of potential free energy (between 2 transition states)
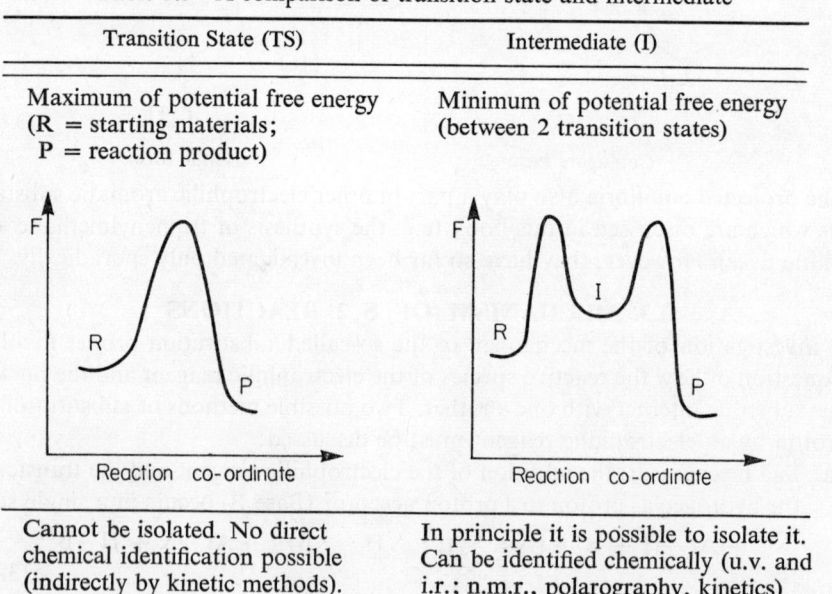	
Cannot be isolated. No direct chemical identification possible (indirectly by kinetic methods).	In principle it is possible to isolate it. Can be identified chemically (u.v. and i.r.; n.m.r., polarography, kinetics)

So far with electrophilic substitutions it has been possible to establish experimentally only the S_E2 mechanism. Here it is seen that the large differences observed in the course of electrophilic reactions are attributable not to a change of the mechanism but to the large differences in the relative free-energy levels of R, TS_1, I, TS_2 and P (Table 3.3). The substitution proper has been most rigorously investigated in relation to azo coupling[10] and to the structure of the intermediate products in nitrations, Friedel–Crafts reactions and halogenations.

In cases where the amount of intermediate product does not increase during the reaction (3.13)–(3.14) the so called steady-state equation of M. Bodenstein (3.15) is valid for the S_E2 substitution. This applies in the following conditions: $k_1 \ll k_2$ [B] and/or $k_1 \ll k_{-1}$.

$$-\frac{d[X^\oplus]}{dt} = [X^\oplus][Ar\text{—}H] \frac{k_1(k_2[B]/k_{-1})}{1 + (k_2[B]/k_{-1})} \qquad (3.15)$$

When, furthermore, $k_2[B] \gg k_{-1}$, it is seen that the quotient of the rate constants in (3.15) is approximately equal to k_1, i.e. independent of k_2, k_{-1} and [B]. Conversely the reaction rate becomes proportional to the base concentration [B] and the quotient k_2/k_{-1}, when $k_2[B] \ll k_{-1}$.

Thus it follows that there must be electrophilic aromatic substitutions that are base catalysed and those that are not affected by the concentration and type of the added bases as long as we can disregard the preequilibria (section 3.2).

Both extreme cases ($k_2[B] \gg k_{-1}$ or $k_2[B] \ll k_{-1}$), as well as their transitions, have been found with azo couplings.[10] The reaction of 4-chlorobenzenediazonium ion with 1-naphthol-4-sulphonic acid is not base catalysed ($k_2[B] \gg k_{-1}$), whilst that of 1-naphthol-3-sulphonic acid, or 2-naphthol-8-mono- and 6,8-disulphonic acid is moderately or strongly base catalysed ($k_2[B] \sim k_{-1}$ or $k_2[B] \ll k_{-1}$).

The cause of the different ratios of the reaction rates in these reactions is to be found in the steric conditions of the intermediate. In the intermediate with 2-naphthol-8-mono- or 6,8-disulphonic acid strong steric hindrance exists between the bulky sulphonic group in the peri position to the entering azo nitrogen and to the base B that serves as proton acceptor ($k_2[B] \ll k_{-1}$); the diazo group is therefore readily split off. In the intermediate product of coupling with the 1,3-isomer this hindrance is smaller in the 2-, as well as in the 4-, position since the sulphonic group here is only in the o-position; in the 1,4-isomeric intermediate product the sulphonic group and the diazonium residue, which now occupies the 2-position, do not obstruct one another at all.

As is to be expected, a kinetic hydrogen-isotope effect is found in cases where k_2 belongs to the rate determining part of the substitution reaction. The corresponding naphtholsulphonic acid with a deuterium atom at the reaction site reacts more slowly in these cases than the protium compound.[11] Table 3.4 shows that the numerical value k_H/k_D of the isotope effect rises with increasing steric hindrance in the intermediate product because, as mentioned above, k_{-1} diminishes in this

Table 3.4 Kinetic hydrogen-isotope effects found in azo coupling reactions of *p*-chlorobenzene diazonium ions with naphtholsulphonic acids[11]

Coupling component (X = H or D)	k_H/k_D	Steric hindrance in the intermediate
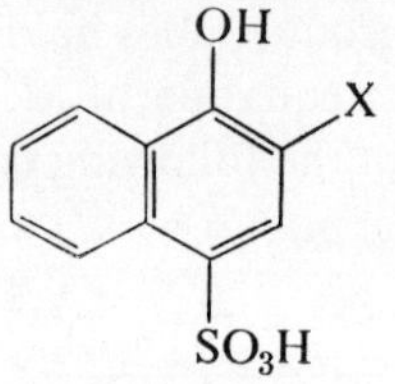 1-hydroxy-naphthalene-4-sulphonic acid (X at position 2)	1·04	small
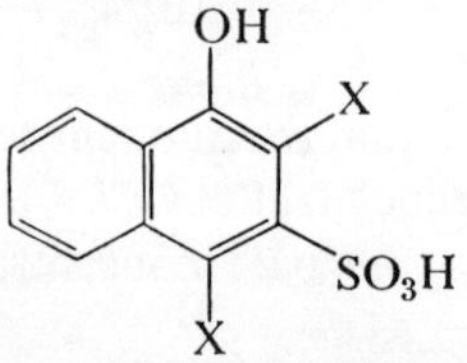 1-hydroxy-naphthalene-2-sulphonic acid (X at positions 3 and 4)	3·10	medium
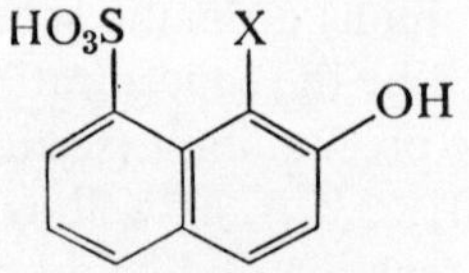 2-hydroxy-naphthalene-8-sulphonic acid (X at position 1)	6·2	large
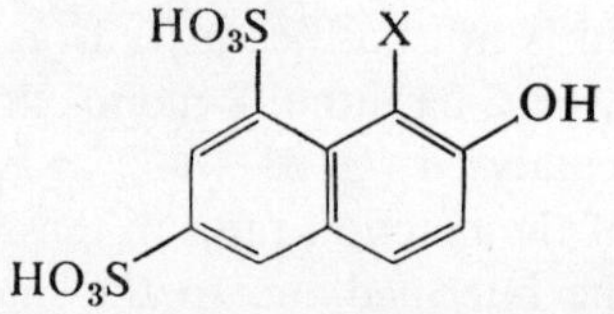 2-hydroxy-naphthalene-6,8-disulphonic acid (X at position 1)	6·55	large

sequence. Similarly, it can also be demonstrated that the size of the isotope effect can be influenced by variations in k_2 and in [B].

All proton acceptors that are present in the system can serve as base B in equation (3.15); they influence rate-limiting proton transfers. The addition of bases influences the reaction rate even when the pH value (i.e. the hydroxyl ion concentration) remains unchanged. This is, therefore, a typical example of a so-called *general base catalysis*, whilst the effect of the hydroxyl ion concentration on the preequilibria which was dealt with in section 3.2 is characteristic of a *specific base catalysis* (hydroxyl ion catalysis). Here the addition of other bases (without changing the pH) does not influence the preequilibria; it will catalyse the coupling only when k_2 belongs to the rate-determining part of the overall reaction.

With electrophilic substitutions that are subjected to general base catalysis, a choice of suitable bases makes it possible to catalyse the reactant without unfavourably shifting the pre-equilibria equilibria (cf. section 3.2). Pyridine and mixtures of its homologues have proved to be particularly valuable as base in azo coupling since the nitrogen is the nucleophilic centre at the edge of the planar pyridine molecule and is therefore in a position to reach the proton of the intermediate product, which is 'concealed' in a sterically unfavourable environment (cf. section 4.1.3).

In recent years there has been experimental evidence for the existence of two types of intermediates of electrophilic aromatic substitutions, the so-called σ- and π-complexes.[12] The intermediate product of azo coupling of 2-naphthol-6,8-disulphonic acid, which is described above, is termed a σ-complex because the electrophilic reagent and nucleophilic substrate are linked by a σ bond. In contrast, the linkage in a π-complex is due to the fact that the nucleophilic substrate, as electron donor, contributes all its electrons to the electron-deficient electrophilic reagent (electron acceptor). It is still discussed whether a substitution proper first involves the π-complex with the electrophilic reagent as electron acceptor, followed by the σ-complex and finally a π-complex with a proton that is to be transferred as electron acceptor.

The general base catalysis plays a part not only in the general acceleration of electrophilic aromatic substitutions but it also influences the orientation of the substitutions (cf. section 4.1.3).

3.4. NUCLEOPHILIC AROMATIC SUBSTITUTIONS

There are three types of nucleophilic aromatic substitutions[13] (Table 3.1). Which of these types participates in a reaction depends essentially on the kind of nucleofugal leaving group, the presence of activating substituents and the type of the nucleophilic reagent.

By treating diazonium salts in dilute, aqueous solutions at the boil, the diazonium group is replaced by a water molecule. This process is followed by a fast proton release resulting in the formation of the corresponding phenol. Recently E. S. Lewis and coworkers[13a] have indicated that the reaction proceeds not by an $S_N 1$ mechanism but that it is $S_N 2$-like: the elimination of N_2 and addition of H_2O proceed in a (one-step) concerted mechanism (3.16). The diazonium group as molecular nitrogen is an excellent nucleofugal leaving group.

$$\text{Ar}-\text{N}_2^{\oplus} + \text{H}_2\text{O} \rightarrow \text{Ar}-\overset{\oplus}{\text{O}}\text{H}_2 + \text{N}_2$$
$$\downarrow -\text{H}^{\oplus} \qquad (3.16)$$
$$\text{Ar}-\text{OH}$$

By the action of very strong bases (e.g. $NaNH_2$), or in extreme conditions (350–380°C with the Dow phenol process) the benzyne mechanism[14] converts an aryl halide into a phenol or amine via a metastable intermediate with a formal triple bond, the so called benzyne or aryne.

Since two possibilities exist when proton and base add to the arin, mixtures of isomeric substituted aryl halides arise (cf., for example, (3.17) in respect of 4-alkyl halogenobenzene).

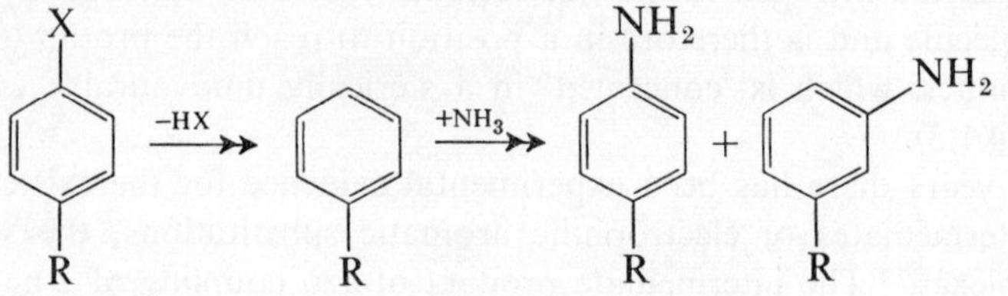

$$\underset{R}{\underset{|}{C_6H_4}}-X \xrightarrow{-HX} \underset{R}{\underset{|}{C_6H_4}} \xrightarrow{+NH_3} \underset{R}{\underset{|}{C_6H_4}}-NH_2 \;+\; \underset{R}{\underset{|}{C_6H_4}}-NH_2 \qquad (3.17)$$

X = Halogen

Nucleophilic aromatic substitutions are most frequently used in accordance with the addition–elimination mechanism elucidated mainly by Bunnett and co-workers.[15] Under the influence of strong electron acceptors A, which serve as potential nucleofugal leaving groups, in the *o*- and/or *p*-position to the substituent X the base (e.g. $OH^\ominus$, NH_3) attaches itself to the carbon which is carried by the leaving group. This gives rise to an adduct that can be isolated in some cases and to which the structure shown in (3.18) is usually assigned.[16]

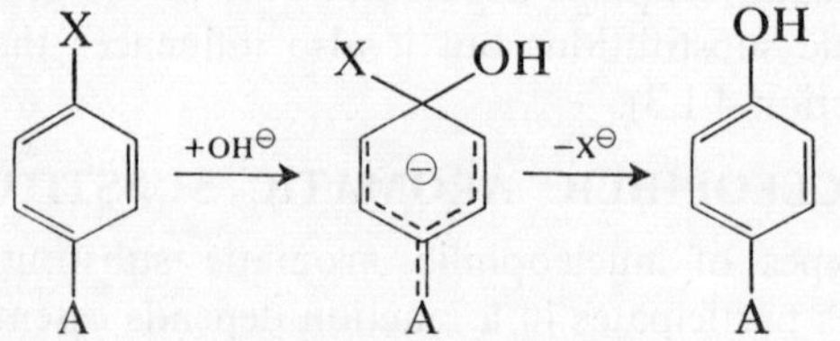

$$\underset{A}{\underset{|}{C_6H_4}}-X \xrightarrow{+OH^\ominus} \underset{A}{\underset{|}{C_6H_4}}\underset{OH}{\overset{X}{<}}^{\ominus} \xrightarrow{-X^\ominus} \underset{A}{\underset{|}{C_6H_4}}-OH \qquad (3.18)$$

X = Halogen, SO_3H
A = NO_2, COAr, COR, CN (also in 2-, 2,4- or 2,4,6 positions)

This type of reaction is of the greatest importance in the production of many dyestuff intermediates (alkali fusions, aminolyses etc), which are not dealt with in this book; it plays an important rôle, usually combined with redox processes, in the synthesis of anthraquinone dyes (cf. sections 10.2 to 10.4).

3.5. THE QUINONE-HYDROQUINONE REDOX SYSTEM

Important parallels exist between redox reactions and heterolytic substitutions, since oxidizing and reducing agents are electron acceptors and donors, respectively. For this reason aromatic compounds with substituents which facilitate electrophilic substitution are readily oxidized, a fact which is frequently utilized in the synthesis of dyestuff intermediates.[1a]

Reactions which, in principle, can be traced back to the system benzoquinone ⇌ hydroquinone (colour photography, Scholl reaction for the production of higher condensed ring systems, the application of vat dyes and others), are of specific significance in dye-synthesis proper.

The Fundamentals of the Reaction Mechanisms of Dyestuff Syntheses 35

The reduction of benzoquinone to hydroquinone is formally a reaction involving two hydrogen atoms. However, it can be demonstrated that this reversible reaction is composed of a system of equilibria where two electrons are transferred *stepwise* from the reducing agent and, as both these reduction processes are coupled with acid–base equilibria (proton transfers),[17] this gives rise to the non-protonated semiquinone, as well as its conjugate mono and dibasic acids, as intermediates. In the diagram (3.19) the redox equilibria are arranged in a horizontal, and the acid–base equilibria in a vertical, direction. Furthermore, in a third dimension, an association (dimerization) equilibrium leads from the radical type of semiquinone to the quinhydrone.

The equilibria depicted in diagram (3.19) for quinone and hydroquinone apply analogously to the *p*-phenylenediamine ⇌ quinoneimine redox system.

The diagram (3.19) shows that quinones have to be good hydride ion acceptors

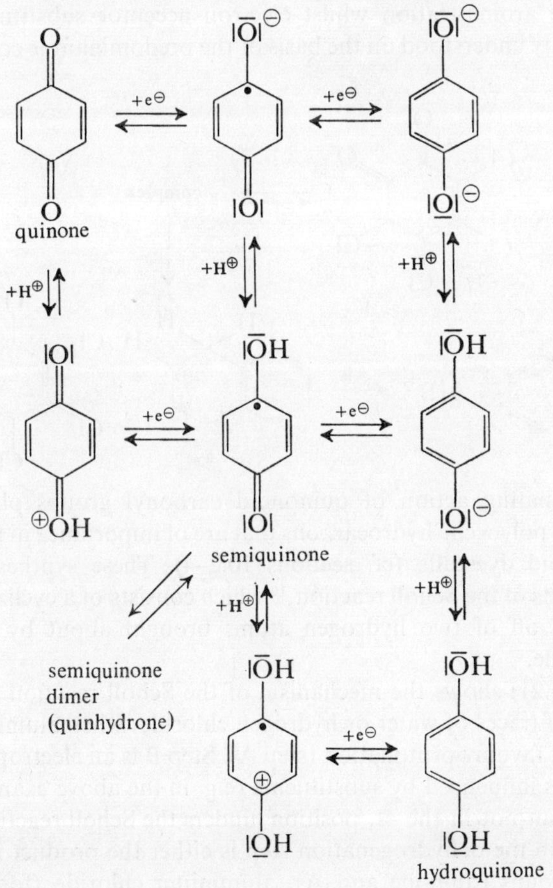

(3.19)

since they are here converted into the monobasic anion of the corresponding hydroquinone, or that the reaction of hydroquinone or *p*-phenylenediamine derivatives with oxidizing agents such as silver ions gives rise to quinones (or quinonediimines) which are electrophilic reagents.

The conversion of 1,2-dihydronaphthalene into naphthalene by means of tetrachloro-1,2-quinone exhibits important characteristics of the action of quinone as proton acceptor in the dehydration of hydroaromatic compounds;[18] dihydronaphthalene as electron donor forms a π-complex with a quinone as electron acceptor (cf. section 3.3), in which the hydride ion transfer proper occurs in a slow reaction step. The carbonium ion which first forms (conjugate acid of naphthalene) subsequently transfers a proton rapidly to a base in the reaction system (3.20). Since this last step does not belong to the rate-determining part of the reaction, the reaction is not generally base catalysed. The reaction velocity is thus influenced by substituents in the aromatic ring of the dihydronaphthalene, in that the electron donors accelerate aromatization whilst electron acceptor substituents retard it. This effect is readily understood on the basis of the predominant π-complex equilibrium.

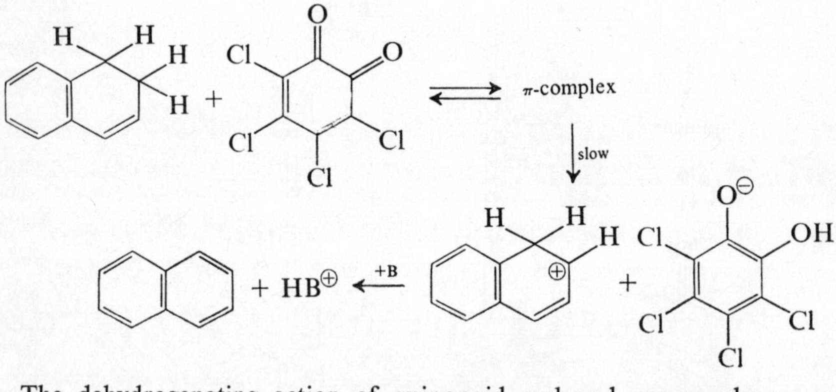

The dehydrogenating action of quinonoid carbonyl groups plays a part in many syntheses of polycyclic hydrocarbons that are of importance in the production of anthraquinonoid dyestuffs (cf. sections 10.2–4). These syntheses can all be regarded as variants of the Scholl reaction,[19] which consists of a cyclization together with the splitting off of two hydrogen atoms brought about by the action of aluminium chloride.

The diagram (3.21) shows the mechanism of the Scholl reaction. The necessity for the presence of traces of water or hydrogen chloride, when aluminium chloride is used as catalyst, favours protonation (step A). Step B is an electrophilic substitution; as usual, it is influenced by substituents (e.g. in the above example a benzoyl group in the 5-, but not in the 4-, position hinders the Scholl reaction). Acting as hydride acceptor in the dehydrogenation step is either the product itself if it contains reducible groups (quinone and A), aluminium chloride (formation of an

Al—H compound analogous to LiAlH$_4$) or additives (O$_2$, MnO$_2$, nitrobenzene, etc.). In cases where the product acts as hydride ion acceptor, air will reoxidize it to quinone in the subsequent process on a preparatory scale. The steps A to C are reversible but the hydride ion transfer (D) is practically irreversible.

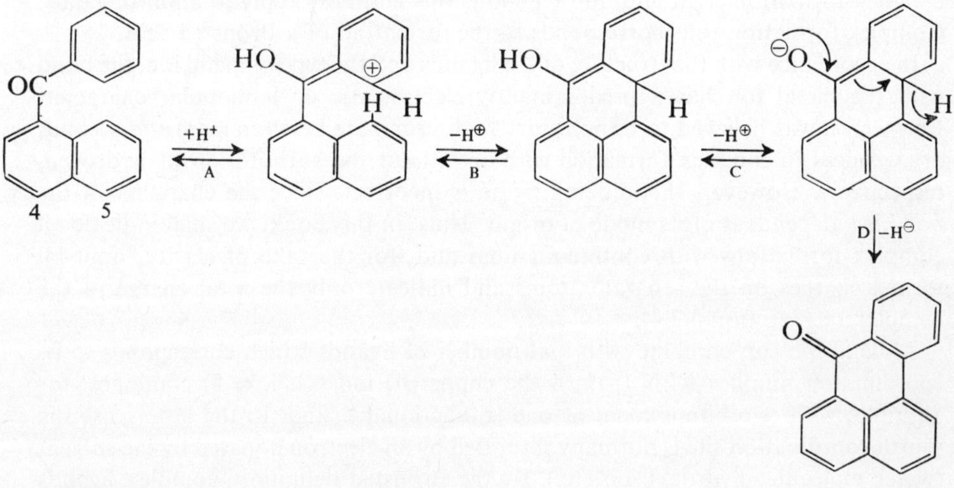

(3.21)

Whilst in the Scholl reaction the hydride ion transfer is preceded by a protonation with subsequent electrophilic substitution, other cyclizations of condensed aromatic ring systems (e.g. indanthrone fusions, carbazole formations, etc.) which are carried out in a strongly alkaline medium (alkali fusion) presumably proceed via primary splitting off of protons and nucleophilic substitution by the resultant carbonium or via a transient primary nucleophilic hydroxyl ion adduct followed by nucleophilic condensation by splitting off of the hydroxyl ion before transfer of the proton occurs (cf. sections 10.2–4).

3.6. EQUILIBRIA IN THE FORMATION OF METAL COMPLEXES

Metal complexes are of great importance not only in nature (chlorophyll, haemin, vitamin B$_{12}$, enzymic catalysis etc.), but also in technical dye chemistry. Nowadays they play a part in different groups of azo dyestuffs, the aza-[18]annulene structures (phthalocyanine, Chapter 7) and in the manufacture of stabilized diazonium salts. Their use as mordant dyes derived from nitroso and anthraquinone compounds now has only historical interest. In contrast, dyes used as indicators in complexometric titrations are still important.[20]

The term metal complex signifies the combination of a metal ion with a ligand, i.e. a particle (ion or molecule) which contains free electron pairs. Bi- or polyfunctional ligands can occupy two or more coordination sites of the metal ion and thus form cyclic complexes (chelates).

The addition of a ligand to a metal ion in solution is a reversible reaction. Here, with hydroxyl, carboxyl and (occasionally) amino groups a hydrogen ion is replaced by a coordination site of the metal ion.

When lone pairs of electrons act as ligand sites with carbonyl oxygen atoms, complex formation is an addition reaction; this can also apply to amino ligands. Complex formation thus corresponds to the formation of a Brönsted acid.

In accordance with the structure of the ligands and the type of metal ion, the bond with the metal ion has a predominantly electrostatic or homopolar character. Formerly it was believed to be necessary to discriminate between main and secondary valences in complex formation with or without the substitution of hydrogen, respectively. However, this differentiation is incorrect, since the character of the bond is independent of its mode of origin. Thus, in this book, we shall indicate all complex formations with continuous lines and, for the sake of clarity, omit all formal charges on the separate atoms and indicate only the total charge of the complex (e.g. section 4.2.4).

A metal ion can combine with that number of ligands which corresponds to its coordination number (C.N.); thus, the copper(II) ion (C.N. = 4) combines, for example, with two bifunctional or one trifunctional ligand. In the latter case the fourth coordination site is normally saturated by an electron donated by the solvent (water molecule, hydroxyl ion etc.). By the Brönsted definition, complex ligands are bases. A metal ion competes with protons in attempting to react with a ligand in aqueous solution. Through the partial loss of basic components, the solution becomes acid during complex formation, and the process can therefore be followed by potentiometric titration in the same way as a neutralization. The equations (3.22), (3.23) for the protolysis, and the complex equilibrium of a metal amine as an example of complex formation without hydrogen substitution, can thus be combined in (3.24).

$$\overset{\oplus}{NH_4} \underset{}{\overset{K_a}{\rightleftarrows}} NH_3 + H^{\oplus} \tag{3.22}$$

$$Me^{n\oplus} + NH_3 \underset{}{\overset{K_{cI}}{\rightleftarrows}} [Me(NH_3)]^{n\oplus} \tag{3.23}$$

$$Me^{n\oplus} + \overset{\oplus}{NH_4} \underset{}{\overset{K_{ca}}{\rightleftarrows}} [Me(NH_3)]^{n\oplus} + H^{\oplus} \tag{3.24}$$

$$K_{cI} = \frac{[Me(NH_3)]^{n\oplus} \cdot [H]^{\oplus}}{[NH_4]^{\oplus} \cdot [Me]^{n\oplus} \cdot K_a} \tag{3.25}$$

In accordance with (3.25) the complex formation constant K_{cI} is a measure of the stability of the complex $[Me(NH_3)]^{n\oplus}$. (3.25) shows that, in this case too, the stability of the complex is a function of the hydrogen ion concentration. Further monofunctional ligands can also be bound in conformity with the C.N. and their equilibrium constants are formulated similarly. The formation of complexes with several ligands therefore corresponds to the neutralization of polybasic Brönsted acids.

Equation (3.25) furthermore demonstrates that the stability of the complex depends on the basicity of the ligands (or the acidity constant K_a of the conjugate acid); more stable complexes are formed by stronger bases. It is, however, possible to increase the stability of the complexes far more effectively by other means; namely by the use of chelates as illustrated by a comparison of the overall formation constants of complexes of nickel with ammonia (forming a hexamine) with that of the hexamino compound (3.1).

$$R = \begin{array}{c} H_2NCH_2CH_2 \\ \diagdown \\ \diagup \\ H_2NCH_2CH_2 \end{array} N-CH_2-CH_2-N \begin{array}{c} CH_2CH_2NH_2 \\ \diagup \\ \diagdown \\ CH_2CH_2NH_2 \end{array}$$

(3.1)

$$Ni^{2\oplus} + 6NH_3 \underset{}{\overset{K_c}{\rightleftarrows}} [Ni(NH_3)_6]^{2\oplus} \qquad K_c = 3\cdot 1 \times 10^8 \qquad (3.26)$$

$$Ni^{2\oplus} + R \underset{}{\overset{K_{c'}}{\rightleftarrows}} [Ni(R)]^{2\oplus} \qquad K_{c'} = 2 \times 10^{19} \qquad (3.27)$$

Although (3.1) is chemically very similar to ammonia, its nickel complex is more stable by about 11 orders of magnitude (3.26), (3.27). This tremendous effect of polyfunctional ligands is of decisive importance in relation to metal complex dyes. It was discovered by Schwarzenbach[21] and termed *chelation effect*.

Complex formation with bi- and polyfunctional complexing reagents corresponds to that with monofunctional ligands. By analogy with the equilibria (3.22), (3.23) or the overall equation (3.24), an increase in pH leads to reaction with more than one ligand site of the complex-forming compound. This effect of pH is utilized in the manufacture of chromium complexes of o,o'-dihydroxyazo dyes that act as trifunctional ligands; according to the pH value during formation, one or two dye particles occupy three or six coordination sites of the chromium ion. Tetrafunctional and hexafunctional complexing dyes are also known (cf. section 4.2.4.).

Apart from formation equilibria the *stereochemistry* of the metal complexes is of importance with dyestuffs. This deals with the possibilities of isomerism arising from different arrangements of the ligand around the metal ion.

Metal ions with C.N. = 4 (e.g. copper (II) and nickel (II) ions) form tetrahedral or planar complexes. Complex tetrahedra with two identical unsymmetrical bifunctional complex formers (A − B + A − B) give rise to a pair of optical isomers, whilst with a planar arrangement geometrical, but no optical, isomers are possible.

Complexes of metal ions having C.N. = 6 (e.g. Cr (III) and Co (III) ions) and bi- and trifunctional unsymmetrical chelating agents are of special interest in the dyestuff field. Three equivalents of a bifunctional compound A−B (e.g. salicylic acid) can occupy all six coordination sites (so-called 1:3 complex). When only two equivalents react, 1:2 complexes form in which the fifth and sixth coordination sites are occupied by the solvent (compare above). Here *cis-* and *trans*-isomers as well as a larger number of optical isomers are possible.

With complexes containing two trifunctional chelating agents A—B—C an all-*trans* form with a centre of symmetry is possible. The planes of the two chelating agents are perpendicular to one another. Pfeiffer[22] was able to demonstrate both the expected optical enantiomers with a 1:2-Cr (III) complex of an o,o'-dihydroxyazo dye. Apart from this type of perpendicular arrangement there are five further possibilities of arranging the two trifunctional chelating agents in two parallel planes at three coordination sites of the chromium octahedron; these three positions are situated at the corners of the same faces of the octahedron. Schetty[23] discovered this type of complex only a few years ago (cf. section 4.2.4).

3.7. LITERATURE

1. A. A. Frost and R. G. Pearson, *Kinetics and Mechanism*, 2nd edition, John Wiley, New York, 1961; E. Gould, *Mechanism and Structure in Organic Chemistry*, H. Holt and Co., New York, 1960; P. Sykes, *Reaktionsmechanismen der organischen Chemie, eine Einführung*, 4th edition, Verlag Chemie, Weinheim, 1970; L. P. Hammett, *Physical Organic Chemistry*, 2nd edition, McGraw-Hill, New York, 1970.
1a. H. Morschel, in *Ullmann's Encyclopädie der technischen Chemie*, Vol. 19, 3rd edition, Urban and Schwarzenberg, München, 1969, pp. 271–418.
2. C. K. Ingold, *Structure and Mechanism in Organic Chemistry*, 2nd edition, Cornell University Press, Ithaka, 1970, Chap. 6. cf. also P. B. D. de la Mare and J. H. Ridd, *Aromatic Substitution (Nitration and Halogenation)*, Butterworth, London, 1959; R. O. C. Norman and R. Taylor, *Electrophilic Substitution in Benzenoid Compounds*, Elsevier, Amsterdam, 1965; J. G. Hoggett, R. B. Moodie, J. R. Penton and K. Schofield, *Nitration and Aromatic Reactivity*, Cambridge University Press, Cambridge, UK, 1971.
3. P. Kovacic and coworkers, *J. Am. chem. Soc.*, **87**, 1262 (1965), **88**, 100, 1000, 3819 (1966).
4. J. H. Ridd, *Quart. Rev. (chem. Soc., London)*, **15**, 418 (1961), *J. Soc. Dyers Colourists*, **81**, 355 (1965) and other publications in *J. chem. Soc. (London)*.
5. C. Wittwer and H. Zollinger, *Helv. chim. Acta*, **37**, 1954 (1954); J. S. Littler, *Trans. Faraday Soc.*, **59**, 2296 (1963).
6. A. Hantzsch, *Ber. dtsch. chem. Ges.*, **27**, 1702 (1894).
7. R. Kübler and W. Lüttke, *Ber. Bunsenges. physik. Chem.*, **67**, 2 (1963).
8. H. Zollinger, *Chemie der Azofarbstoffe*, Birkhäuser, Basel, 1958; H. Zollinger, *Diazo and Azo Chemistry, Aliphatic and Aromatic Compounds*, Interscience, New York, 1961.
9. R. Wistar and P. D. Bartlett, *J. Am. chem. Soc.*, **63**, 413 (1941); H. Zollinger and C. Wittwer, *Helv. chim. Acta*, **35**, 1209 (1952).
10. H. Zollinger and coworkers, *Helv. chim. Acta*, **38**, 1597, 1617, 1623 (1955), **41**, 1816, 2274 (1958).
11. H. Zollinger, *Advances physic. org. Chem.*, **2**, 163 (1964).
12. G. A. Olah and M. W. Meyer, in G. A. Olah, *Friedel–Crafts and Related Reactions*, Interscience, New York, 1963, I, Chap. 8.

13. J. Miller, *Aromatic Nucleophilic Substitution*, Elsevier, Amsterdam, 1969; R. Foster, *Organic Charge Transfer Complexes*, Academic Press, London, 1969; F. Pietra, *Quart. Reviews (Chem. Soc. London)*, **23**, 504 (1969); D. V. Banthorpe, *Chem. Reviews*, **70**, 295 (1970); M. J. Strauss, *Chem. Reviews*, **70**, 667 (1970).
13a. E. S. Lewis, L. D. Hartung and B. M. McKay, *J. Am. Chem. Soc.*, **91**, 419 (1969).
14. R. Huisgen and J. Sauer, *Angew. Chem.*, **72**, 91 (1960); H. Heaney, *Chem. Reviews*, **62**, 81 (1962); R. W. Hoffmann, *Dehydrobenzene and Cycloalkynes*, Verlag Chemie, Weinheim, 1967; E. K. Fields and S. Meyerson, *Advances physic. org. Chem.*, **6**, 1 (1968).
15. J. F. Bunnett, *Quart. Rev. (chem. Soc., London)*, **12**, 1 (1958), as well as other publications, especially in *J. Am. chem. Soc.*
16. P. Caveng, P. B. Fischer, E. Heilbronner, A. L. Miller and H. Zollinger, *Helv. chim. Acta*, **50**, 848, 861, 866 (1967), as well as references therein from V. Gold, K. L. Servis, R. Foster and C. A. Fyfe.
17. L. Michaelis and M. Schubert, *Chem. Reviews*, **22**, 437 (1938).
18. L. M. Jackman and D. T. Thompson, *J. Chem. Soc. (London)*, **1961**, 4794.
19. A. T. Balaban and C. D. Nenitzescu, in G. A. Olah, *Friedel–Crafts and Related Reactions*, Interscience, New York, 1964, II, 979.
20. G. Schwarzenbach and H. Flaschka, *Die komplexometrischen Titrationen*, Enke Verlag, Stuttgart, 1965.
21. G. Schwarzenbach, *Helv. chim. Acta*, **35**, 2344 (1952).
22. P. Pfeiffer and S. Saure, *Ber. dtsch. chem. Ges.*, **74**, 935 (1941).
23. G. Schetty and W. Kuster, *Helv. chim. Acta*, **44**, 2193 (1961), and further publications from G. Schetty (*Helv. chim. Acta*, 1961–1970); R. Grieb and A. Niggli, *Helv. chim. Acta*, **48**, 317 (1965); H. Jaggi, *Helv. chim. Acta*, **51**, 580 (1968).

4

Azo Dyes

Azo dyes are compounds containing azo groups (—N=N—) which are linked to sp^2-hybridized carbon atoms. In accordance with the number of such groups the dyes are described as mono-, dis-, tris-, tetrakis- (etc.) azo dyes. The azo groups are mainly bound to benzene or naphthalene rings, but in some cases they are also attached to aromatic heterocycles (e.g. pyrazole) or enolizable aliphatic groups (e.g. acetoacetic acid derivatives).

The azo compounds represent the largest group of industrial dyes, both in number and amount produced.

4.1. PRINCIPLES UNDERLYING THE PREPARATION OF AZO DYES

Among the methods for producing aromatic azo compounds, azo coupling, i.e. the reaction of an aromatic diazo compound with a so-called coupling component, is of outstanding significance. The second and third sections of this Chapter will be devoted to its discussion. In the first section other modes of formation of aromatic azo compounds are summarized.

4.1.1. General methods

The parent substance, azobenzene, was obtained by Mitscherlich,[1] nearly a quarter of a century before the discovery of diazo compounds, by the action of alcoholic caustic potash on nitrobenzene. Nitrobenzene is reduced to azobenzene more conveniently by tin or iron in caustic alkaline solution, glucose and tin (II) salts, as well as by electrolysis.

Under certain conditions, the action of reducing agents on diazonium salts does not lead, as might be expected, to hydrazines, but to azobenzene derivatives. For this purpose, the action of ammoniacal copper oxide on diazonium salts[2] is particularly suitable. According to Saunders and Waters,[3] aryl radicals, which are presumably formed initially, react with a diazonium ion in the presence of the reducing agent.

Symmetrical azo compounds are also formed by the introduction of diazonium groups through diazo group transfer.[2a]

The oxidation of amines occasionally leads to aromatic azo compounds.[4a]

Methods which produce unsymmetrical derivatives are more important than the above to an azo dyes' chemist.

The azo synthesis by Suckfüll and Dittmer[5] makes it possible to produce azo compounds which do not carry any electron donor groups (OH, NH_2 etc.). The synthesis consists of the reaction of a diazonium ion with a diazosulphonate (4.1). However, a mechanistic investigation[6] has shown that the range of preparative applications of this reaction is limited.

$$Ar-N_2^{\oplus} + Ar'-N_2-SO_3^{\ominus} \xrightarrow[\text{in } H_2O]{} Ar-N=N-Ar' + N_2 + H_2SO_4 \quad (4.1)$$

Zincke and Bindewald[7] found that 4-phenylazo-1-naphthol, produced by coupling diazobenzene with 1-naphthol, is also obtained by the reaction of phenylhydrazine with 1,4-naphthoquinone. This method is of importance for the production of some hydroxyazo compounds that are not readily, or not at all, accessible via azo coupling (for instance, 2-phenylazo-1-naphthol) as well as for the interpretation of the hydroxyazo-quinonehydrazone-tautomerism (section 4.2.1.).

The reaction of nitroso compounds with amines can give rise to azo derivatives. However the application of this reaction is limited to benzene derivatives.

The thermolysis of amines with nitrobenzene derivatives and caustic soda, according to Martynoff,[8] results, in a manner which is not quite clear, in a good yield of azo compounds.

Oxidative azo coupling discovered by Hünig[9] is very interesting; by this method many azo derivatives of heterocyclic compounds have become accessible. Here amidazone-hydrazones are converted oxidatively by aromatic heterocycles into a diazonium intermediate which, being electrophilic, reacts with a coupling component (e.g. 2-naphthol) to form, with the loss of two hydrogen atoms, the azo compound (4.2).

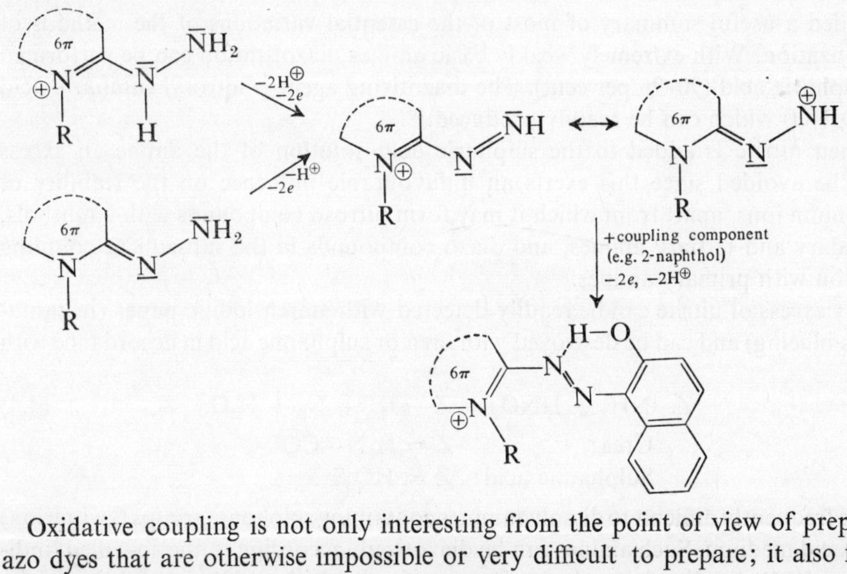

Oxidative coupling is not only interesting from the point of view of preparing azo dyes that are otherwise impossible or very difficult to prepare; it also forms an important systematic link with the synthesis of quinoneimines (Chapter 8).

Furthermore it leads, via the azidinium salts, of which Balli[10] has made a special study, to azamethines (e.g. to pentazamethines (Chapter 6)) and to aliphatic azo compounds. These azidinium salts are also one of the numerous reagents for preparing diazo compounds by means of diazo group transfer.[2a]

4.1.2. Diazotization

The diazotization of aromatic primary amines is the first of two reaction steps by which practically all technical azo dyes are produced. Normally an aqueous solution of the amine is converted into the diazonium ion at a temperature of about 0°C by the action of sodium nitrite in the presence of mineral acid (4.3). The use of at least two (usually 2.5) equivalents of mineral acid in accordance with (4.3) is essential for smooth reaction; this is a consequence of the equilibria that exist (section 3.2).

$$Ar-NH_2 + 2HX + NaNO_2 \rightarrow Ar-N_2^{\oplus}X^{\ominus} + NaX + 2H_2O \qquad (4.3)$$
$$(X = Cl, Br, NO_3, HSO_4 \text{ etc.})$$

A higher proportion of hydrogen ions than that indicated in equation (4.3) is used in diazotizing weakly basic amines since this results in the equilibria in the nitrous acid solution shifting towards the production of the more electrophilic particles so that the amine–ammonium equilibrium does not lie too far towards the non-reactive ammonium ion. Thus, p-nitraniline is dissolved in hot 5–10 N HCl and the solution is either cooled rapidly or poured on ice. Nitranilinium chloride precipitates before hydrolysis to the base occurs. Smooth diazotization results on the immediate addition of nitrite. Fierz and Blangey[4b] described the details and provided a useful summary of most of the essential variations of the methods of diazotization. With extremely weakly basic amines diazotization can be performed in sulphuric acid (90–96 per cent). The diazotizing agent is nitrosyl sulphuric acid (HSO_4NO) which can be readily produced.[4c]

When nitrite is added to the sulphuric acid solution of the amine an excess must be avoided since this exerts an unfavourable influence on the stability of diazonium ions, apart from which it may form nitroso compounds with naphthols, secondary and tertiary amines, and diazo compounds in the subsequent coupling reaction with primary amines.

Any excess of nitrite can be readily detected with starch iodide paper (instantaneous blueing) and can be destroyed with urea or sulphamic acid in accordance with (4.4).

$$Z-NH_2 + HNO_2 \rightarrow Z-OH + N_2 + H_2O \qquad (4.4)$$

Urea: $Z = H_2N-CO-$
Sulphamic acid: $Z = HO_3S-$

It is frequently difficult to dissolve amines containing sulphonic groups (zwitterion) in an acid medium. Such amines can be diazotized, according to the so-called 'indirect method', by dissolving the amine as anion with the addition of the requisite amount of alkali and adding sodium nitrite to the almost neutral solution; this mixture is then added to mineral acid with stirring.

Further methods of diazotization are described by Saunders[11] and Zollinger.[12a,13a]

The mechanism of the diazotization reaction was elucidated by Hughes, Ingold and Ridd[14] (summaries: references 12b, 13b, 15). Nitrosation of the amino group is the essential step in diazotizing. With secondary aliphatic or aromatic amines the reaction stops at the nitroso stage (formation of nitrosamines RR'N—NO). With primary amines the nitrosamine is also formed at first but is quickly transformed (presumably via the diazo hydroxide) into the diazonium ion (4.5).

$$R-NH_2 + \underset{Y}{N=O} \xrightarrow[\text{slow}]{-Y^\ominus} R-\underset{H}{\overset{H}{N^\oplus}}-N=O \xrightarrow{-H^\oplus} R-\underset{\downarrow \text{rapid}}{\overset{H}{N}}-N=O$$

$$R-OH + H^\oplus \underset{\text{rapid}}{\xleftarrow{+H_2O}} R^\oplus + N_2 \underset{\substack{(R=Aryl)\\ \text{very slow}}}{\overset{(R=Alkyl)}{\underset{\text{rapid}}{\xleftarrow{}}}} R-N\equiv N \underset{\substack{-H_2O\\ \text{rapid}}}{\xleftarrow{+H^\oplus}} R-N_2-OH$$

(4.5)†

Aromatic diazonium ions are relatively stable in solution. In contrast, solid diazonium salts are explosive (section 4.2.3). Sources of danger are not only dry diazonium salts but also those diazotizations in which one works at high concentrations, so that suspensions of diazonium salts exist.

The significance of the preequilibria that exist for both reaction components was discussed in section 3.2. From equations (3.1)–(3.3) it can be seen that there is an optimum pH region in diazotizations, dependent on the basicity of the amine and the type of acid that is being used: weakly basic amines are, as already mentioned above, diazotized in moderately, or highly, concentrated acids.

Diazotization is also used for the estimation of aromatic amines; an aqueous acid solution of the amine is titrated with a $NaNO_2$ solution. Here, the addition of KBr accelerates the course of the titration markedly, presumably by reacting with nitrous acid to form nitrosyl bromide (NOBr), which is a good nitrosating agent.

4.1.3. Azo coupling

As already mentioned in section 4.1.1, practically all azo dyes are produced from diazo compounds and coupling components by means of the azo coupling reaction. Since the mechanism of azo coupling has already been dealt with (sections 3.2 and 3.3) only the preparatory and technological consequences resulting from the investigation of the mechanism will be discussed here.

Coupling reactions must be carried out in a medium in which the equilibria of the diazo and the coupling components lie as far as possible towards the diazonium ion, the phenolate anion, the enolate anion or the free amine, depending

† Arrows with 2 points denote two rapid steps.

on whether coupling is performed with a phenol (naphthol, etc.), an enol (acetoacetanilide, 3-methyl-1-phenyl-5-pyrazalone etc.), or with an aromatic amine. This results in an optimum coupling-pH region (section 3.2) for each combination of diazo- and coupling-components. It is limited by those acidities (expressed in pH units) which numerically correspond to the pK value of the coupling- or diazo-rearrangement-constant respectively. This region lies, approximately, between pH 4 and pH 9 with aromatic amines as coupling components, between pH 7 and pH 9 for enols and pH 9 for phenols (pH 9–12 with o-diazophenols as diazo components).

The pH dependence of the coupling rate also explains why the site of reaction with coupling components carrying both aromatic amino and hydroxyl groups may be determined by means of the pH of the reaction medium. With aminonaphthols, e.g. 2-amino-8-naphthol-6-sulphonic acid (γ-acid), coupling in the acid region takes place at the 1-position, i.e. at the aromatic ring which carries the amino group. On the other hand, coupling in strongly-alkaline media takes place almost exclusively at the ring which carries the hydroxyl group (5- and 7-positions).

Raising the temperature does not, in most cases, exert a favourable influence because the diazo decomposition reactions have larger activation energies and therefore a larger temperature gradient than the coupling reactions. Whereas the reaction rate of coupling increases by a factor of 2·0 to 2·4 for every 10°C, that of the decomposition reaction increases by a factor of from 3·1 to 5·3. Moreover, with couplings which, for any reason, have to be performed at such high pH values that an appreciable proportion of the diazo compound is present as diazotate, an increase in temperature is not favourable because it shifts the *cis-trans* rearrangement (section 3.2) almost irreversibly in favour of the *trans*-compound.

With some technical azo couplings the addition of common salt *before* coupling gives a higher yield. This can be attributed to the different dependence of the reaction rate of coupling and diazo decomposition on ionic strength (Brönsted's salt effects).

A number of technically important azo dyes can be satisfactorily produced only by an addition of pyridine or pyridine homologues. These are, in the main, couplings of diazotized aminoazo- and aminodisazo-compounds, bisdiazotized† benzidines, and others required for the synthesis of polyazo dyes. Pyridine is also used for some couplings with diazotized o-aminophenols in order to build up complex-forming dyes (section 4.2.4).

This supplies incontrovertible evidence that pyridine acts by catalysing the splitting-off of the proton from the intermediate product in the substitution step proper. The pyridine molecule is introduced as the proton acceptor B in the second step of the reaction system (3.13)–(3.14) of section 3.3.[16] The kinetic equation (3.15),

---

† For the synthesis of a compound with two diazo groups from a diamine, we use the expression *bisdiazotization*. In the past this process was called tetrazotization, but this term really should be reserved for the introduction of tetrazo groups $-N\!\!=\!\!N-\overset{\oplus}{N}\!\!\equiv\!\!N$.

which was discussed previously, shows that changing the concentration of B exerts an influence on the kinetics of the overall reaction only when the back reaction of the first step (k_{-1}) is faster than proton release of the intermediate product (k_2). This condition is obeyed not only in the presence of bulky substituents in the o- or periposition to the coupling site but also if the diazonium ion is acting as an extremely weak electrophile (e.g. with diazotized o-aminophenols), thereby shifting the ratio $k_2[B]/k_{-1}$ to smaller values. Following this kind of reasoning, one can readily predict whether a particular coupling is likely to be catalysed by the addition of pyridine.

The mode of action of other additives which are recommended in the patent literature is not established with certainty. Urea has a de-associating effect in polyazo coupling.

The general base catalysis underlying this use of pyridine does not only play a role in the general acceleration of azo coupling but it also influences the *orientation*. This effect is principally of importance in coupling derivatives of 1-naphthol-3-sulphonic acid since azo dyes which are interesting from a technical point of view can be obtained from these compounds only when coupling occurs in the 2- (not 4-) position (sections 4.2.1 and 4.2.5). With these coupling components the 2- as well as the 4-position is generally base catalysed. The yield ratio of o- to p-coupling product corresponds to the ratio of the measured velocity constants k_o/k_p. It can be determined experimentally that the coupling velocity constant k_p is increased more strongly by bases than k_o. The product ratio thus shifts in favour of the (undesirable) p-product by raising the base concentration. The greater catalysis of the reaction at the p-position can be understood as shown in Table 4.1:

Table 4.1 Inductive effect and steric hindrance in the intermediate of azo coupling of 1-naphthol-3-sulphonic acid in the 2- and 4-positions, respectively[17]

Intermediate of substitution (Az = Arylazo)	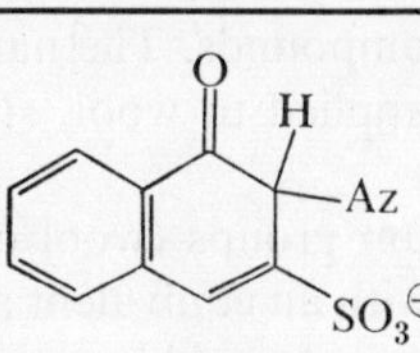	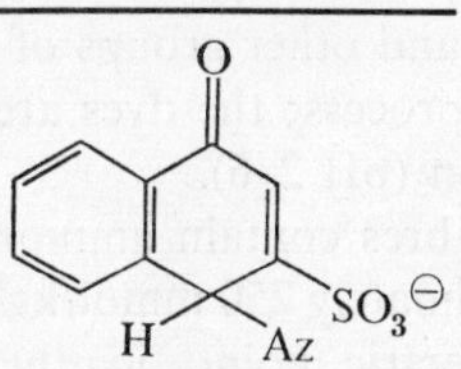
Steric hindrance at the reaction site k_{-1}	small small	large large
-I-Effect k_2	large large	small small
$k_2[B]/k_{-1}$	very large	very small
Base catalysis	weak	strong

the acidifying effect of the carbonyl group on k_2 affects the adjacent o-position more strongly than the p-position, whilst the p-position is impeded to a greater extent by the adjacent peri-hydrogen.[17]

In this sense the o/p ratio with a particular coupling component depends on the structure of the diazo component, in that the more strongly electrophilic diazonium ions favour the p-position. For this reason it is here frequently difficult to obtain a satisfactory yield of the o-product. When coupling with diazotized aminophenols, very high pH values are, therefore, employed: this shifts the proportion of the diazophenol, itself present only in a small equilibrium concentration, towards the less reactive diazophenolate zwitterion, which is more strongly inclined to attack the o-position.

This and other special features of the technically important coupling reactions have been summarized.[18]

4.2. APPLICATION IN DYEING

4.2.1. Anionic monoazo dyes

The presence of one, or more, water solubilizing ionizable substituents is characteristic of anionic monoazo dyes. They consist almost exclusively of sulphonic groups because these are readily accessible and because they are strong electrolytes and therefore almost completely dissociated in all activity ranges occurring in aqueous solutions.

These dyes are widely used for paper and leather and their heavy metal salts (Ba, Ca etc.) are used as pigments. Their main application, as so-called acid dyes, is for dyeing protein and synthetic polyamide fibres.

The term 'acid dye' denotes a large group of anionic dyes with a relatively low molecular weight that carry 1 to 3 sulphonic groups. They are mainly monoazo compounds, but also include disazo, nitro, 1-amino-anthraquinone, triphenyl-carbonium and other groups of compounds. The name 'acid dye' is derived from the dyeing process; the dyes are applied to wool, silk and polyamides in weakly acid solution (pH 2–6).

Protein fibres contain ammonium groups (wool contains about 850 mmol/kg, silk approximately 250 mmol/kg) with an equivalent amount of carboxylate groups at the isolectric point. Synthetic polyamides presumably exist as zwitterions in water, and as amino carboxylic acids, with 30–50 mmol/kg amino groups and 50–70 mmol/kg carboxyl end groups when dry.

In dyeing with acid dyes, a more or less large fraction of carboxylate groups is neutralized in accordance with the pH value, and a gegenion (chloride, hydrogen sulphate, acetate, etc.) is adsorbed, thus preserving electrical neutrality. The true dyeing process consists of an *ion exchange* of the adsorbed gegenions for dyestuff anions, because the affinity of these dye anions is greater than that of the (small) gegenions of the acid used. The greater the affinity of the dyestuff anions for the substrate, the smaller is the hydrogen ion concentration of the dyebath necessary to ensure the adsorption of the dye. For this reason the highly wash resistant 'milling' dyes, with strong affinity, are dyed in more weakly acid solution than ordinary acid dyes. The physical-chemical basis of dyeing equilibria and the nature of the forces of affinity are discussed in section 11.3.1. The adsorption of metal complex dyes (section 4.2.4) is based on analogous mechanisms.

Aniline derivatives mainly serve as diazo components. 5-pyrazolone derivatives are commonly used as nucleophilic substrates in coupling to produce yellow dyes while naphthol- and naphthylamine-derivatives are used for orange to bluish violet dyes. Here it is important that the hydroxyl or amino groups of the coupling components are in the *o*-position to the azo bridge since in this case a hydrogen bond can form between the hydroxyl or amino group and the azo nitrogen in the *β*-position. Consequently the acid–base equilibria of the hydroxyl or amino groups are displaced into pH regions which are outside the practical applicability of these dyes (pH <2 and >11). This is of practical significance because dissociation of the hydroxyl group or protonation of the amino group leads to undesirable colour changes.

The colourist aspect of these equilibria can be demonstrated by the example of isomeric dyes Naphthalene Orange I (C.I. Acid Orange 20) (**4.1**) and Naphthalene Orange G (C.I. Acid Orange 7) (**4.2**). In contrast to Orange G, Orange I is little used now because the tint changes in soda- and washing-tests. The pK_2 values show that Orange I is already present as a dibasic anion above pH 8·2 while Orange G first splits off its phenolic proton in a pH range not attainable in washing.

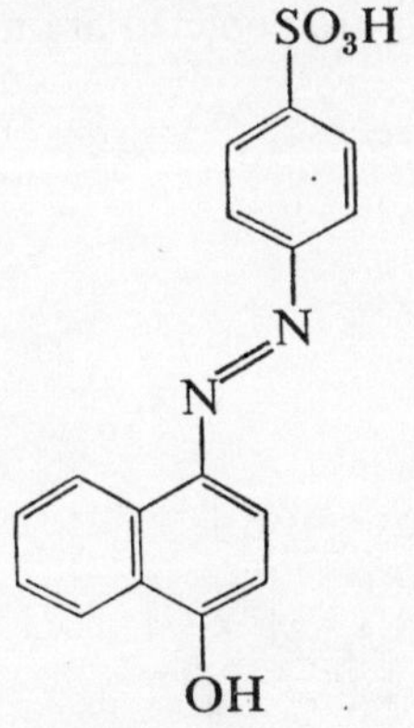

(**4.1**) pK_2 = 8·2 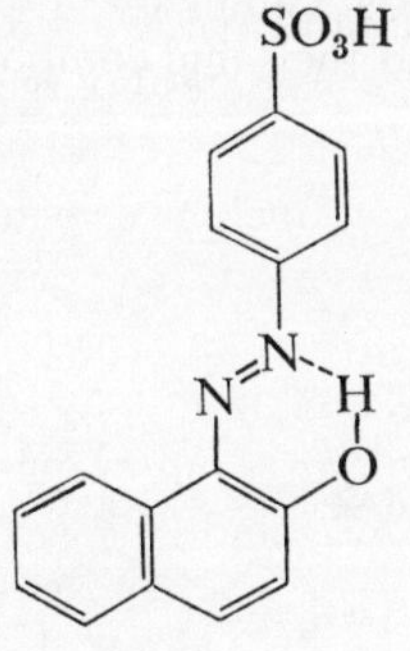   (**4.2**) pK_2 = 11·4

In cases where it is not possible to force coupling in the *o*-position to the OH- or NH$_2$-group, it is necessary to carry out subsequent alkylation or acylation. With coupling components which carry an additional second amino or hydroxyl group it is also necessary, in most cases, to inactivate them by means of acylation or tosylation in order to obtain dyes which have good alkali- and acid-fastness. Azo Geranine 2G (C.I. Acid Red 1) **(4.3)** from 1-acetylamino-8-naphthol-3,6-disulphonic acid (acetyl-H-acid) constitutes an important example in this respect. The corresponding compound which has a free amino group is technically valueless.

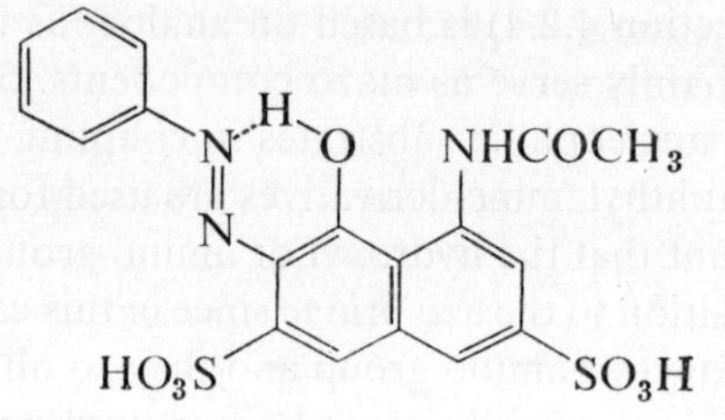

(4.3)

The replacement of the acetyl group by other acyl residues changes the colouristic properties of this type of dye in a characteristic manner. The water solubility of the corresponding dyes decreases and the affinity for protein fibres increases in the sequence acetyl > propionyl > n-butyryl > benzoyl ~ p-tosyl > p-tertiary-butylbenzoyl ~ capryl ($C_7H_{15}CO$—). The washing fastness increases in the above order. With all azo compounds having OH-, NH_2- or NHR-groups there is a tautomeric equilibrium between the azo- and the hydrazone-form. Equation (4.6) gives the example of the coupling products of the 1-naphthol (e.g. Naphthalene Orange I), and shows that both tautomeric forms are interconnected via *one* conjugate base. (Both structures of the conjugate base which are depicted are mesomeric limiting structures of one and the same compound.)

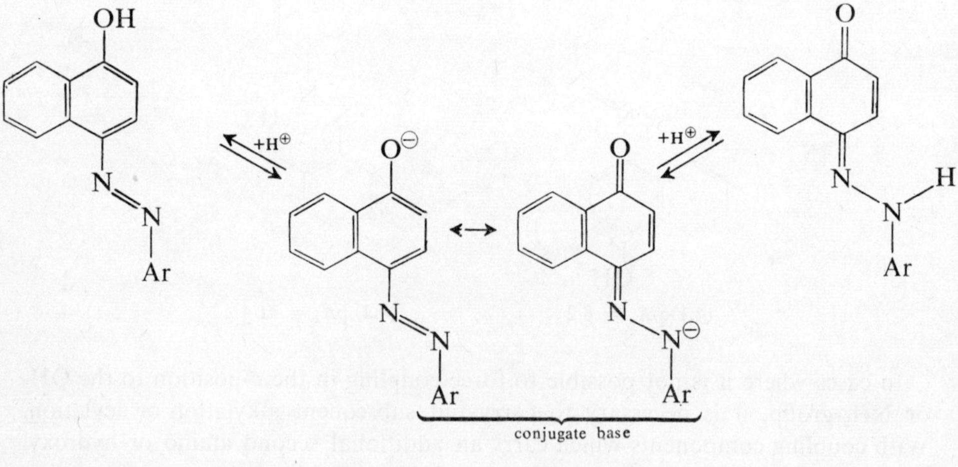

(4.6)

Because of the rapidly established equilibrium (4.6) it is also possible, in principle, to produce such dyes from hydrazine and quinone. The tautomeric equilibrium with phenylazophenols favours the hydroxyazo-, with phenylazonaphthols, the ketohydrazone-, and with phenylazonaphthylamines the aminoazo-form.[19] Insufficient is known about the state of equilibrium of technical dyes; in this book we have written the azo, not the hydrazone structure, except when a tautomeric form is known, e.g. the metal complexes (4.8) and (4.9).

Chemically, many of the reactive dyes (section 4.2.6) belong to the group of anionic monoazo dyes; these contain, apart from the usual structural features, a residue which is capable of condensation with hydroxyl groups of cellulose (cotton), and with the amino and mercapto groups of protein fibres (wool).

4.2.2. Disperse azo dyes

Disperse dyes are almost insoluble in water; they can be applied to cellulose acetate fibres as well as most fully synthetic fibres from aqueous suspensions. Most yellow, orange and red disperse dyes are azobenzene derivatives. In contrast to anionic monoazo dyes, benzene, not naphthalene, derivatives are the most important coupling components. In this respect, the N-(2-hydroxyethyl)- and N-2(-methoxyethyl)anilines are particularly significant because these compounds are not completely insoluble in water but possess the low water solubility necessary for the dyeing process (section 10.5.2). The use of nitrodiazobenzenes as diazo components is much more frequent with these dyes than with water-soluble dyes (e.g. Celliton Scarlet B (C.I. Disperse Red 1) (4.4)). Great caution is needed when diazotizing di- and trinitro derivatives of aniline because of the danger of explosion at high temperatures.

$$O_2N-\underset{}{\bigcirc}-N\underset{N}{\overset{}{=}}-\underset{}{\bigcirc}-N\underset{CH_2-CH_2-OH}{\overset{C_2H_5}{<}}$$

(4.4)

A recent publication shows the multiplicity of the constitutions of modern disperse dyes and the relation between constitution and colouring properties.[20] In 1968, 50 per cent of disperse dyes used technologically were monoazo-, 25 per cent were anthraquinone-, and 10 per cent were disazo-compounds.

In recent years disperse dyes have gained importance through the use of aromatic heterocyclic compounds as diazo or coupling components. As electrophilic reactants the diazonium salts of nuclear-substituted 2-aminothiazoles, 2-aminobenzthiazoles, 2-aminoisothiazoles, 5-aminopyrazoles, and 2-aminothiazoles are employed. Besides these, 5-pyrazoles, 2-methyl- and 2-phenylindole, 1,3,3-trimethyl-2-methylenindolenin, imidazole and pyridone are used as coupling components. Barbituric acid as well as hydroxyl-containing quinolines and quinolones are also used.

4.2.3. Azoic dyes

Azoic dyes are water-insoluble mono-, and (a few) disazo-, dyes for cellulose fibres, which are formed *on the fibre* by coupling a water-soluble diazo compound with a water-soluble coupling component having affinity for cellulose.[21] The decisive step in the development of this class of dye was the discovery of a coupling component that is soluble in water as an anion and has affinity for cellulose, the so-called Naphtol AS (2-hydroxy-3-naphthanilide). This coupling component is combined on the fibre with diazobenzene or its substitution products. Since the resulting azo dyes do not carry any sulphonic groups and the hydroxyl group of the coupling component is in the *o*-position to the arylazo residue (its acidity is thus markedly weakened because of the hydrogen bond (section 4.2.1)), these dyes are almost insoluble in water and therefore possess very good wet fastness.

Apart from Naphtol AS proper, a large number of derivatives substituted at the phenyl nucleus are commercially available. These produce only orange to blue colours. Therefore, further water-soluble coupling components with an affinity for cellulose were developed which are not naphthols in the chemical sense but are nevertheless termed 'Naphtols', e.g. Naphtol AS–G, from which yellow acetoacetanilide-disazo dyes are obtained (**4.5**).

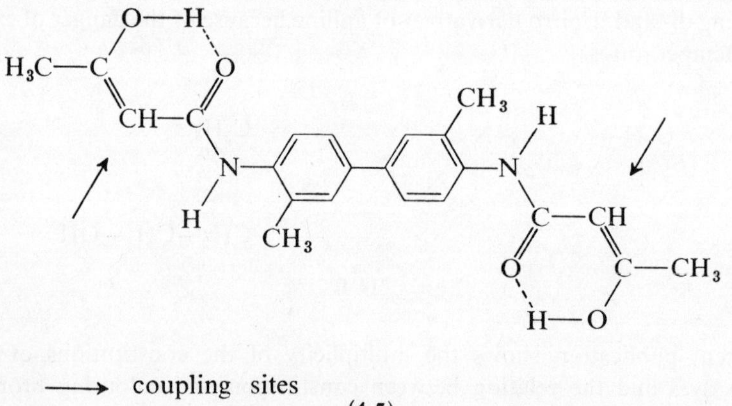

⟶ coupling sites

(**4.5**)

Derivatives of 4-aminodiphenylamine, e.g. Variamine Blue Base B (4-amino-4'-methoxydiphenylamine), are used as diazo components for blue colours. For black colours, either condensed ring systems (carbazol, dibenzofuran etc.) serve as coupling components, or aminoazo compounds serve as diazo components (e.g. Fast Black Base K = 4-amino-2,5-dimethoxy-4'-nitroazobenzene). Since diazotization of some important diazo components (the so-called fast colour bases) at a dyeing mill is awkward, the following further developments of the Naphtol AS dyes have been worked out.

1. *Stabilized diazo salts*

It is not possible to isolate large quantities of diazonium salts in the pure form since they are explosive (section 4.1.2). Diazonium salts in the form of zinc chloride double salts can be stable enough to be isolated, dried and stored ('fast colour salts'). The sulphonic acids of benzene and naphthalene derivatives are also suitable as stabilizers. In this case stabilization depends on the formation of electron donor-acceptor complexes; the aromatic system of the sulphonic acids acts as an electron donor to the (electrophilic) diazonium cation.[22]

2. *Rapid Fast dyes*

The diazo compounds are converted into *trans*-diazotates (section 3.2) and are mixed with Naphtol AS compounds. Alkaline aqueous solutions of these mixtures are applied to the fibre and then 'acid developed'. At higher temperatures, the addition of acid shifts the diazo equilibrium towards the diazonium ion which then rapidly couples.

3. *Rapidogen dyes*

A reaction with primary or secondary aliphatic or aromatic amines changes the diazonium ions into the corresponding diazoamino compounds which are again mixed with Naphtol AS compounds. Acid development on the fibre converts the diazoamino compounds into diazonium ions. The course of the reaction (4.7) shows that this is an example of a specific hydrogen ion catalysis. The Rapidogens have almost completely displaced the Rapid Fast dyes because it is possible to choose the amines so that all their diazoamino compounds, in contrast to the diazotates, have approximately the same dissociation rate at a definite hydrogen ion concentration and temperature in accordance with equation (4.7). Thus, combinations can be chosen that can be developed even in neutral steam, but the solutions of which do not deteriorate, by azo coupling, in a short time at room temperature (Neutrogene-, Rapidogen N-dyes etc.).

$$\text{Ar—N}{=}\text{N}{\diagdown}^{R}_{R} + H^{\oplus} \rightleftarrows \text{Ar—N}{=}\overset{H}{\underset{\underset{R}{\oplus}}{\text{N—N—R}}} \rightarrow \text{Ar—N}_2^{\oplus} + \text{HN}{\diagdown}^{R}_{R} \qquad (4.7)$$

4. *Rapidazol dyes*

In some cases where the *trans*-diazotates are not stable the corresponding *trans*-diazosulphonates (section 3.2) are used.

Naphtol AS combinations produced as solids are also important as pigments.

Since ordinary Naphtol AS combinations frequently have too low a solvent fastness—it is true that they are almost completely insoluble in water but not in some other solvents—it is necessary to raise their molecular weights. This can be done by producing the corresponding primary disazo dyes by coupling bifunctional Naphtol AS compounds with diazonium salts, or bisdiazo compounds with monofunctional Naphtol AS derivatives. Since in the conventional performance of this coupling reaction the second step is not completed for reasons of solubility, it is necessary to couple, for example, bisdiazotized benzidine, not with Naphtol AS itself, but with 2-hydroxy-3-naphthoic acid; under certain conditions it is then possible to convert both carboxyl groups into the acid chlorides which, on subsequent reaction with aniline derivatives, give the corresponding anilides. Cromophtal dyes are made on this principle.[22]

4.2.4. Complex-forming monoazo dyes

Metal complexes of monoazo compounds are principally used as chromium and cobalt complexes for dyeing protein and polyamide fibres as high-grade solvent dyes and as lakes. Copper complexes of polyazo compounds have the character of direct dyes and will be dealt with in this connection later (section 4.2.5.). Complex-forming dyes are predominantly trifunctional, but are, in some cases, also bi-, tetra-, and hexafunctional (section 3.6).[23]

Bifunctional complex-forming dyes contain salicyclic acid as the coupling component, e.g. Alizarine Yellow R (C.I. Mordant Orange 1), in which *p*-nitrodiazobenzene is the diazo component. After-treatment on the fibre with bichromate results in a 1:2-complex of tervalent chromium in which the two residual coordination sites are available for bonding to those groups of the fibre which possess lone-pair electrons (**4.6**, B = —NH_2, —OH, also H_2O). These linkages are the cause of the increased wet fastness of such complexes. For chroming on the fibre (under certain conditions) it is also possible to use chromic salts. Bichromate, which is used industrially, initially oxidizes parts of the fibre and is then converted into chromic ions.

(4.6)

Azo dyes which carry substituents with lone pairs of electrons (—OH, —NH_2, —COOH) in both positions *ortho* to the azo bridge act as trifunctional complex formers. Among these the *o,o′*-dihydroxyazo compounds are the most important. One of the two azo nitrogen atoms serves as the third complex ligand. The nitrogen atom that acts as ligand has been determined, by means of X-rays, in only two cases.[24]

Molecular models, however, show unequivocally that only one of the two atoms is involved in complex formation. Notations with a hyphen or an arrow from the centre of the azo double bond to the metal ion are therefore incorrect.

Schetty[24a] recently showed that in metal complexes of o,o'-dihydroxyazo dyes with the structure A—N=N—B (A and B being aromatic or heteroaromatic systems with hydroxyl groups in the α-positions to the azo bridge) only the nitrogen atom adjacent to the less nucleophilic system acts as the ligand. For example, in 1-(2'-hydroxyphenylazo)-2-naphthol the ligand is the nitrogen adjacent to the 2'-hydroxyphenyl group. Furthermore Schetty showed that the 6-membered chelating ring which is formed with the more nucleophilic system has a quinonehydrazone and not a hydroxyazo structure. This is in agreement with what is known about the position of the hydroxyazo-quinonehydrazone tautomeric equilibrium of the non-metalized hydroxyazo compounds as well as with the analogy between the Brönsted acids and the metal complexes (section 3.6).

With metal ions of coordination number (C.N.) 4 [e.g. Cu (II)] the fourth position is generally occupied by a solvent molecule. Since such copper complexes, even when they contain one or two sulphonic groups, are, in many cases, very difficult to dissolve in water, the possibility that the fourth coordination site is occupied by the second azo nitrogen atom of a further complex-forming dye molecule has been discussed. This would give rise to polymeric layer complexes.

Ternary ligands with ions of C.N. 6 [e.g. Cr (III), Co (III)] give 1:1 and 1:2 complexes (1 metal ion to 1 or 2 dye molecules, respectively); with the 1:1 complexes the three free coordination sites of the metal ion are occupied by water or hydroxyl ions. It is a general rule that 1:1 complexes are prepared in an acid medium (pH < 4), whereas 1:2 complexes are prepared in weakly acid to alkaline media. The stability of the finished dyes is analogous; 1:1 complexes must be dyed in mineral acid, 1:2 complexes in weakly acid or neutral media because disproportionation occurs at high acidity. The overall equations (4.8) and (4.9) show that these phenomena are a consequence of the dependence of the complex equilibria on hydrogen ion concentration (section 3.6).

Chroming on the fibre presumably gives rise to 1:2 complexes. Eriochrome Black T (C.I. Mordant Black 11) (**4.7**), Neolan Blue 2G (C.I. Acid Blue 158) (**4.8**) and Irgalan Grey BL (C.I. Acid Black 58) (**4.9**) are characteristic examples

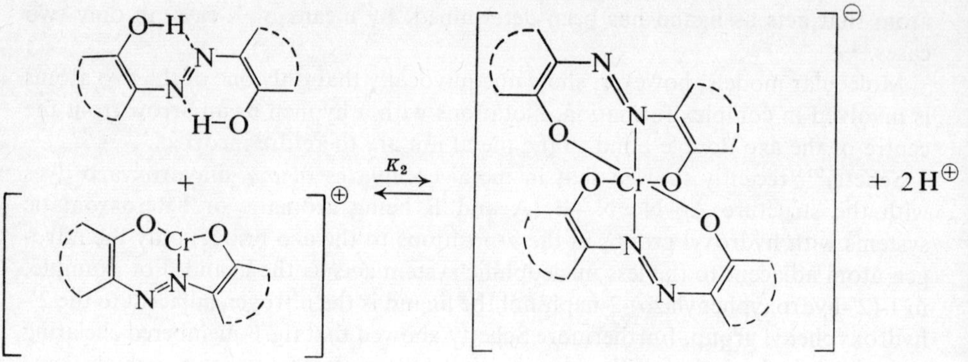

(4.9)

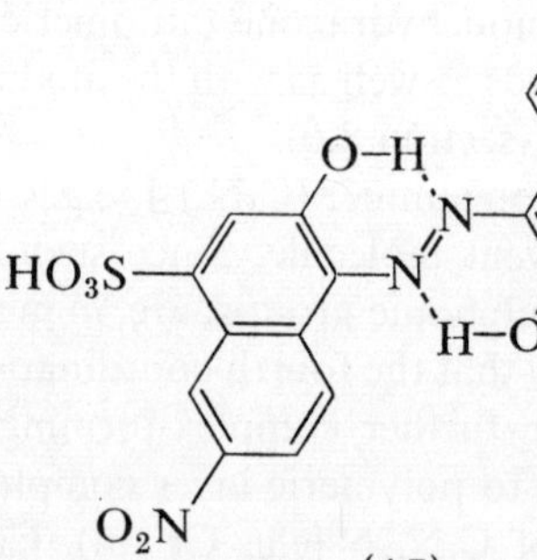

(4.7)

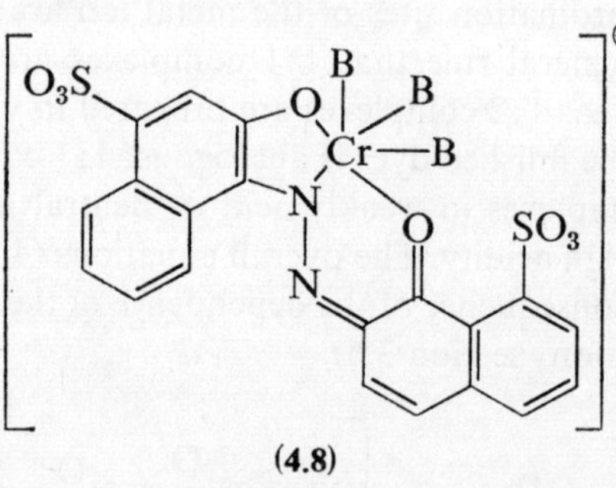

(4.8)

of dyes which can be chromed on the fibre in the above-mentioned manner and of the 1:1 and the 1:2 types of the so-called premetallized dyes, respectively.

Eriochrome Black T is not only an important chrome dye for wool but it also serves as indicator for complexometric titrations. It is prepared from 6-nitro-1-diazo-2-naphthol-4-sulphonic acid (nitro-diazoxy acid) and 2-naphthol.

(**4.8**) and (**4.9**) are termed premetallized dyes because, in contrast to chrome dyes (e.g. **4.7**), the metal is introduced during manufacture, at the final stage of synthesis.

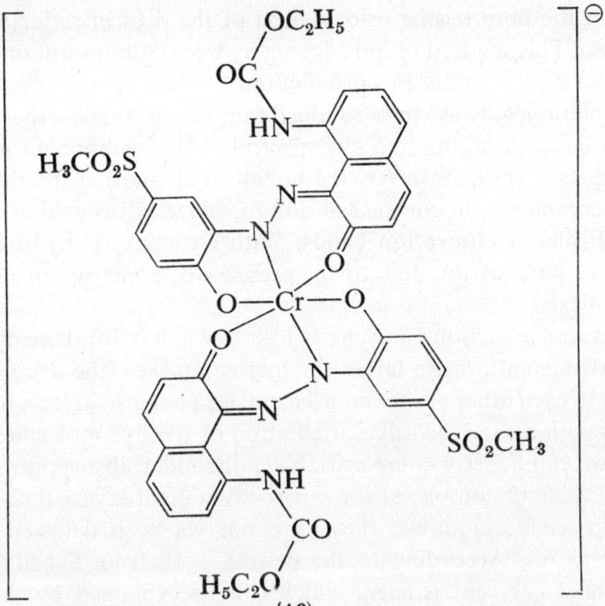

Azo Dyes

(4.9)

Whilst the 1:1 complexes always carry one or two sulphonic groups, the 1:2 complexes do not contain any ionized substituents; but, as **(4.9)** shows, the dye as a whole is an anion. The presence of unionized hydrophilic groups is, on the contrary, of importance. Apart from alkylsulphone residues **(4.9)**, sulphonamido- and, in certain cases, acylamido-groups have been used for these dyes.

With tetrafunctional complex forming dyes, in place of the one hydroxyl group in dyes of the o,o'-dihydroxyazo type, a glycollic acid group (OCH_2COOH, cf. section 4.2.5), or sulphonamide groups with anthranilic acid as the amino component are found; in the latter, the amido nitrogen and one oxygen atom of the carboxyl group serve as the site of the ligand.

As shown by Schetty,[25] it is possible to obtain hexafunctional complex formers by condensing α,ω-diamines with two equivalents of o-nitrobenzene sulphonic chloride and reducing both nitro groups to amino groups. This compound is then bisdiazotized and coupled with two equivalents of 2-naphthol. Formally, the dyes obtained with Cr (III) and Co (III) are 1:1 complexes, but are chemically of the 1:2 type because all six coordination sites are occupied by the ligand groups of the dye; they are pentacyclic complex compounds. As is to be expected, because of the chelation effect (section 3.6), they are very stable and give dyeings having excellent fastness to light and washing.

Some of the older chrome dyes are apparently only bifunctional, since they carry only one hydroxyl group in the o-position. In these cases however the action of bichromate on the fibre results in oxidation of the dye, introducing a second o'-hydroxyl group. This method of introducing hydroxyl groups not only is performed on the fibre but also is utilized in production.

Metal-complex dyes always have subdued (not bright) shades; their visible spectra exhibit broad bands resulting in a low spectral colour component p_e (section 2.2). Chrome complexes (not, however, the cobalt analogues), with the exception of salicylic acid complexes, in contrast to non-metallized dyes exhibit distinct bathochromically displaced absorption bands. With o-carboxy-o'-hydroxyazo dyes the dull shade is in part attributable to the presence of a mixture of different stereoisomeric complexes.

As was discussed in section 3.6, Schetty has shown that, for these dyes, there exists not only the two enantiomeric Drew–Pfitzner complexes (the dye molecules being perpendicular to each other at the central metal ion) but also at least three additional stereoisomers with a sandwich-like orientation of the dye molecules. These structures, which we call Schetty complexes, have dissimilar absorption spectra.

While these Schetty complexes for o-carboxy-o'-hydroxyazo-1:2-complexes are formed in appreciable amounts, they have not yet been detected for o,o'-dihydroxyazo complexes. According to the newest work from Schetty[24a] these have a maximum of 3 per cent isomers which can be explained by the existence of a coordinated sp^3-hybridized azo nitrogen atom. This atom exists as a valence pyramid with a high inversion barrier. Therefore, the dye molecules are slightly bent at this N atom. Because of this, the dye molecules in 1:2 complexes can exist in three different conformations. In addition, the hydroxyazo form which is present in small amounts gives rise to further isomers.

o,o'-Dihydroxyazo dyes have been found to have an interesting application for the spectral sensitization of the semiconductor properties of zinc oxide in electrophotographic reproduction.[25a]

4.2.5. Direct dyes

Compounds which are able to dye cellulose fibres (cotton, viscose, etc.) without the aid of mordants (tannin) are called direct or substantive dyes. The leuco forms of sulphur and vat dyes, the coupling components of the azoic dyes (Naphtol AS) and anionic azo compounds[26] have a substantive character. Anionic monoazo dyes have some affinity for cellulose which is, however, insufficient for practical requirements. The affinity can be increased by enlarging the planar conjugated double bond system (see section 11.3.1). This is done most simply by the introduction of further azo groups, resulting in dis-, tris-, and higher polyazo dyes. Substantive monoazo dyes from derivatives of 2-(4'-aminophenyl-)-5-methylbenzthiazole (dehydrothiotoluidine) and its sulphonation products are of no practical importance nowadays.

With disazo dyes it is necessary to distinguish between primary and secondary types. The former are produced by reacting two equivalents of a diazo component with one bifunctional coupling partner or by the combination of a bisdiazo compound with two equivalents of a coupling component. The monofunctional reagent can be either two equivalents of *one* compound or, in stepwise application, one equivalent each of two different partners.

The secondary disazo dyes result from an azo compound which also carries an amino group which can be diazotized. A disazo dye is produced by diazotization and reaction with a further coupling component. In cases where this second coupling component again carries a free amino group a trisazo dye can be built up.

The most important group of primary disazo dyes are the derivatives of bisdiazotized benzidine and their derivatives which carry two methyl or methoxy groups each in the *o*- and *o'*-position to the amino groups (*o*-tolidine or *o*-dianisidine respectively). Congo Red (**4.10**) was, historically, the first direct dye to be known as such. Bisdiazotized benzidine can readily couple 'unsymmetrically' since the reactivity of the second diazo group is markedly reduced after the first coupling. Reaction with salicylic acid and subsequently with 2-amino-8-naphthol-6-sulphonic acid (γ-acid) produces Diamine Fast Red F (C.I. Direct Red 1).

Substituents in the *m*- and *m'*-position to the amino groups of benzidine lessen the affinity towards cellulose because the coplanar position of both phenyl rings of benzidine becomes impossible. However, when two phenylamino residues are not linked directly, as with benzidine, but via groups which maintain the coplanarity of both phenyl rings, this gives rise to diamines which also produce bisdiazo components for substantive dyes, as is the case with derivatives of the 4,4'-diaminostilbene-2,2'-disulphonic acid. Chrysophenine G (C.I. Direct Yellow 12) (**4.11**), produced by coupling with phenol and subsequent etherification of the phenolic hydroxyl groups with ethyl chloride, is worthy of mention because the coupling product has an indicator character *before* etherification (bright yellow, colour change towards brown at pH 7–8). The cause lies, as with Naphthalene Orange I (section 4.2.1), in the fact that phenols coupled in the *p*-position to the hydroxyl group contain hydroxyl groups which dissociate too readily for purposes of colouration ($pK_a = 7$–8).

Bridges such as —CH_2, —O— and —S— between two phenylamino residues do not give rise to coplanarity of the two phenyl residues; therefore, they are only suitable for the production of anionic disazo dyes for dyeing protein fibres.

(**4.10**)

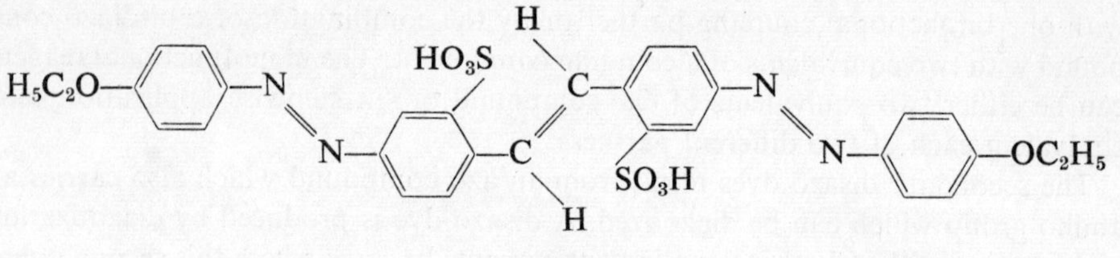

(4.11)

The most important coupling component for secondary dis- and higher polyazo dyes is 2-amino-5-naphthol-7-sulphonic acid (I-acid) and its derivatives in which the amino group has been changed, e.g. N-aryl- and N-aroyl-I-acid (**4.12**) and (**4.13**) imidazolyl, thiazolyl-(**4.14**) and triazolyl-I-acid (**4.15**), as well as derivatives containing two I-acid residues linked via the amino group (Scarlet acid **4.16**, RW-acid **4.17**).

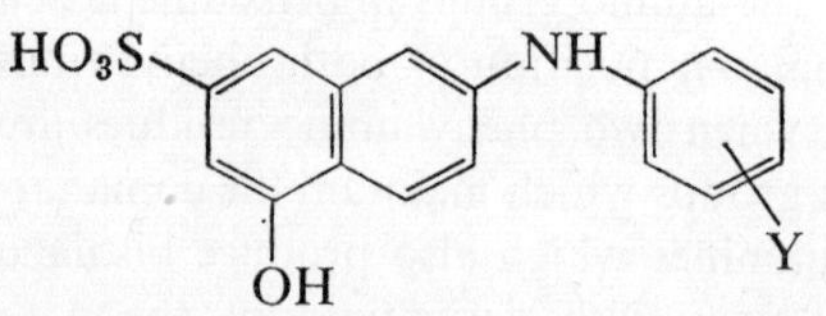

(**4.12**)

Y = H or NH_2 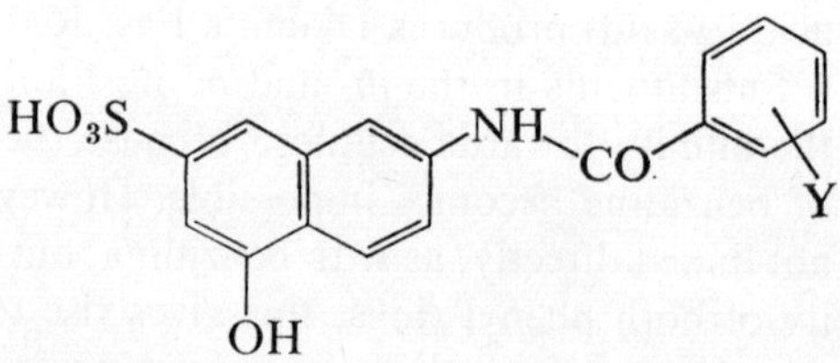   (**4.13**)

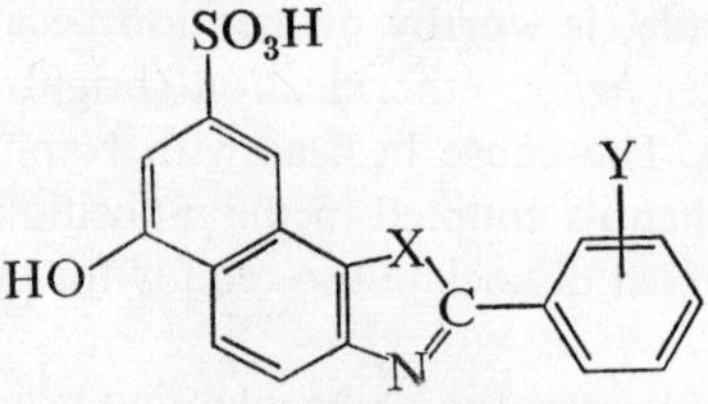

(**4.14**)   X = NH or S   (**4.15**)

Y = H or NH_2

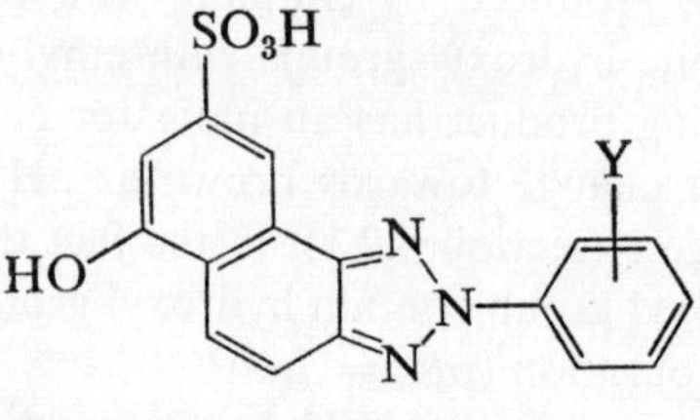

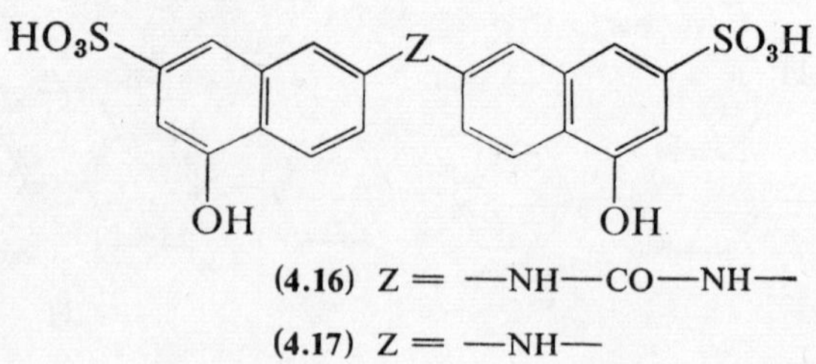

(**4.16**) Z = —NH—CO—NH—

(**4.17**) Z = —NH—

Azo Dyes

I-acid and its derivatives mainly serve as alkaline end components, i.e. those coupled in the 6-position. 1-naphthylamine and its 6- or 7-sulphonic acid are of importance as central components, e.g. Sirius Light Blue BRR (C.I. Direct Blue 71) (2-naphthylamine-4,8-disulphonic acid → 1-naphthylamine → 1,7-naphthylamine sulphonic acid → I-acid). With the 'dimeric' I-acid derivatives (**4.16**) and (**4.17**), coupling occurs with two equivalents in both naphthalene rings (6- and 6′-position). In this way it is relatively easy to build up tetrakis- and hexakis azo dyes such as, for instance, Chlorantine Fast Red 5 BRL (C.I. Direct Red 80) obtained by reacting two equivalents of 4-diazo-azobenzene-2,4′-disulphonic acid with one equivalent of ureido-I-acid (**4.16**). This dye can also be obtained by treating the disazo compound, 4-aminoazobenzene-2,4′-disulphonic acid → I-acid, with phosgene. In contrast to the linkage with phosgene, cyanuric chloride (2,4,6-trichloro-1,3,5-triazine), as bridging member, makes it possible to combine two or three different coloured compounds in the one molecule. By combining yellow and blue dyes in this way, bright fast-to-light green dyes were prepared for the first time, e.g. Chlorantine Light Green BLL (C.I. Direct Green 26) (**4.18**).

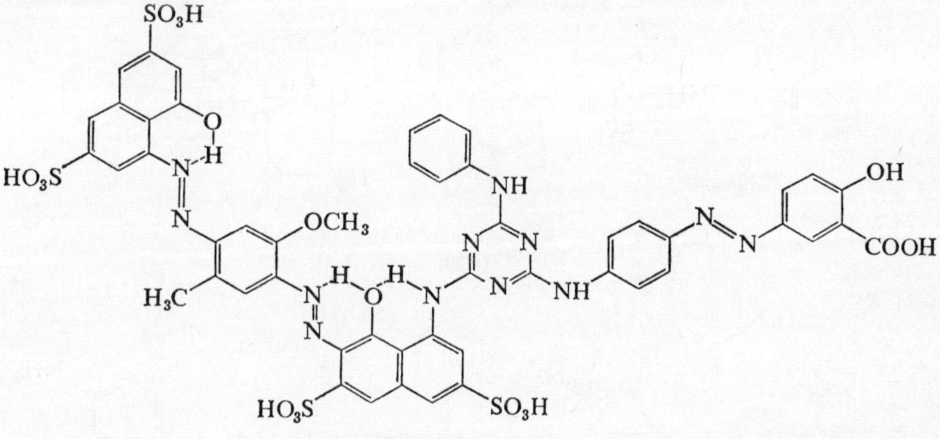

(4.18)

Dis- and polyazo dyes which are transformed into copper complexes either in dye manufacture or on the fibre, carrying tri- or tetrafunctional complex-forming groups, e.g. Coprantine Violet BLL (C.I. Direct Violet 82) (**4.19**) or Benzo Fast Copper Red GGL (C.I. Direct Red 180) (**4.20**) generally exhibit better light-and-wet-fastness than with ordinary direct dyes.

The affinity for cellulose, and thus the wet fastness, can also be improved by again diazotizing the dye that has been adsorbed by the fibre and reacting it with a simple component (e.g. 2-naphthol). This of course presupposes the presence of amino groups which can be diazotized, as in Zambesi Black V (C.I. Direct Black 78) **(4.21)**.

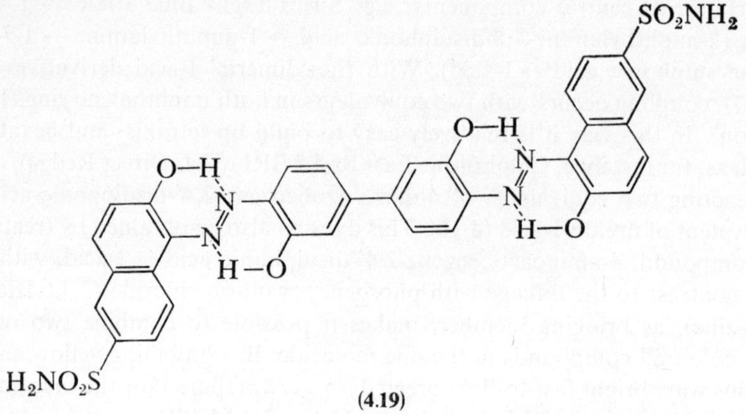

(4.19)

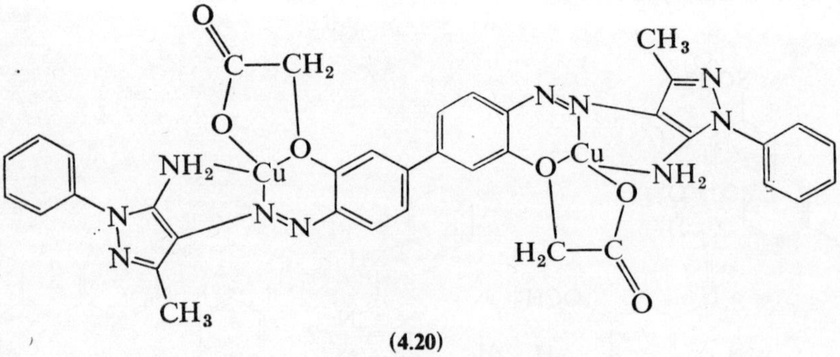

(4.20)

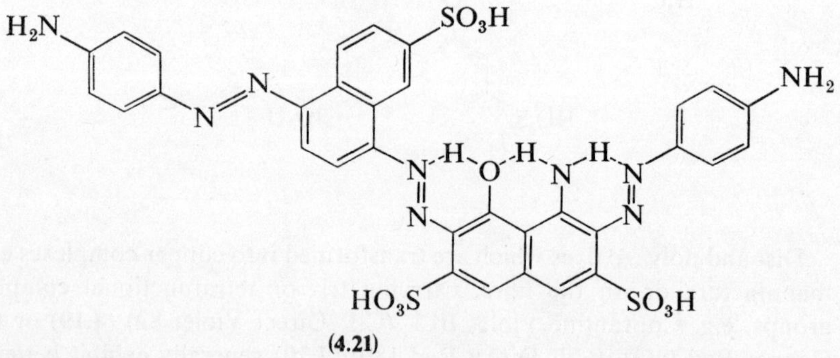

(4.21)

4.2.6. Reactive dyes

Reactive dyes are coloured compounds that react with textile fibres to form covalent bonds. Only a few of the numerous chemical reactions that could serve the purpose have been utilized industrially. This is not surprising, considering the conditions in which reactive dyeing has to be carried out.

1. The fundamental problem of reactive dyeing is that reaction of the reactive dye with water (hydrolysis) competes with the fixation reaction (formation of a covalent bond between the dye and the textile substrate). The hydrolysed dye cannot react with the fibre. A high ratio of fixation to hydrolysis is therefore an important prerequisite for high fixation and therefore for the practical usefulness of a reactive dye.

2. The standard affinity (or substantivity; cf. section 11.3.1) of reactive dyes has to be adjusted to the conditions of application; it must be neither too high, or uniform penetration of the fibres and washing-off of unfixed dye may be difficult, nor too low, as it will have an unfavourable effect on fixation.

3. The wash resistance of reactive dyes depends on the stability of the fibre-dye linkage; the resistance to alkaline or acid hydrolysis of reactive dyeings is closely connected with the degree of fixation because the bonds formed in the alkaline or acid fixation reaction can be hydrolysed in a slower subsequent reaction. Hence, for a reactive dye to be useful, the rate of hydrolysis of dye-fibre bonds must be negligible compared with the fixation velocity. The resistance to alkaline hydrolysis gives information on the stability of reactive dyeings to alkaline washing whilst the resistance to acid hydrolysis is of practical importance in the resin finishing of reactive-dyed cellulose fabrics.

The characteristic structural features of a reactive dye are shown schematically in (**4.22**).

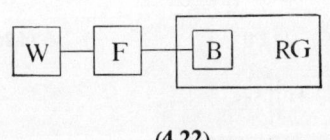

RG: reactive group
B: bridge link
F: chromophore
W: water solubilizing groups

(**4.22**)

Reactive group: The functional groups of fibres, that react with reactive dyes to form covalent bonds, are the hydroxyl group of cellulose and the amino, carboxyl, hydroxyl and thiol groups of wool and silk. With polyamides there are only a few amino end groups and, possibly, carboxyl end groups available. All these residues have a nucleophilic character and therefore add to an electrophilic centre of the reactive group. Several papers have been published on the distribution of the functional groups of the fibre that participate in the fixation process.[27]

According to H. C. Brown[28] a compound behaves the more selectively towards different reactive groups the lower their reactivity. Accordingly, less reactive groups (i.e. groups with low fixation rate) should give higher fixation. Less reactive systems, however, require higher dyeing temperatures, which affect fixation or selectivity (ratio of fixation to hydrolysis) unfavourably owing to the relation (4.10) and the lowering of the standard affinity (cf. section 11.3.1). The effective concentration of adsorbed dye in the fixation reaction is reduced by this change of affinity. The unfavourable influence of high temperature on fixation can, however, be counterbalanced by the choice of a chromophore with greater substantivity.

$$\log \frac{k_{\text{fix}}}{k_{\text{Hy}}} = -k' \cdot \frac{1}{T}(E_{\text{fix}} - E_{\text{Hy}}) \qquad (4.10)$$

k_{fix} and k_{Hy}: rate constants of the fixation or the hydrolysis
E_{fix} and E_{Hy}: activation energy of fixation or hydrolysis
k': proportionality factor

Even with the most appropriate chromophore, the less reactive systems give poorer fixation at higher temperature than very reactive systems at low temperatures.

The reactive groups can be classified according to their reaction mechanisms, thus:

I. Groups that react by the nucleophilic bimolecular (heteroaromatic) substitution mechanism (4.11):
– specific base catalysed addition of the nucleophilic functional group of the textile fibre to the electrophilic centre of the reactive group (4.11: k_1).
– elimination of a nucleofugal leaving group (4.11: k_3); if (4.25) rapidly accumulates, it can be generally base catalysed.[39]

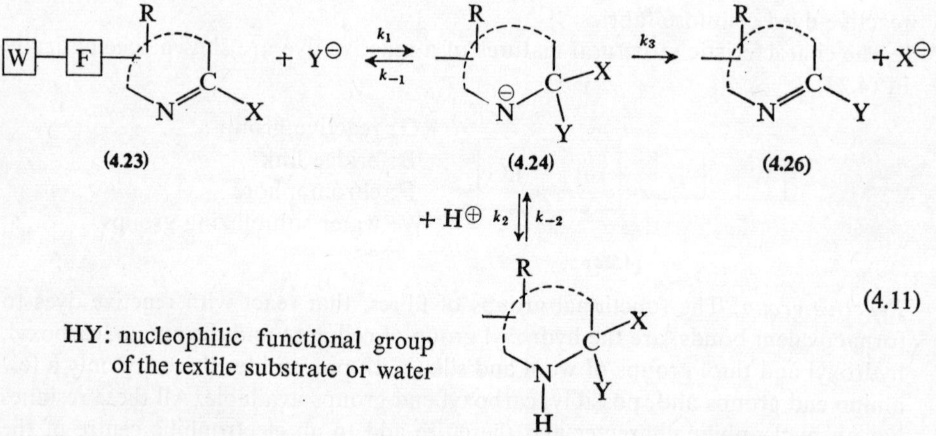

HY: nucleophilic functional group of the textile substrate or water

(4.11)

The following are important representatives of these reactive groups:

di- or monochlorotriazine (**4.27**: Procion M or **4.28**: Cibacron and Procion H dyes),

2,4,5-trihalogenopyrimidine (**4.29a**: Reactone, Drimarene X dyes; **4.29b**: Verofix, Reactolan, Drimalan F dyes),

2,3-dichloroquinoxaline (**4.30**: Levafix E, Cavalite dyes),

2-chlorobenzthiazole (**4.31**: Reatex dyes),

4,5-dichloropyridazone derivatives (**4.32**: Primazine P dyes).

X in (4.11) is a chlorine atom in most of these reactive systems, but also some other nucleofugal leaving groups, such as fluorine, alkyl sulphonic group or quaternary ammonium groups, have to be taken into account.

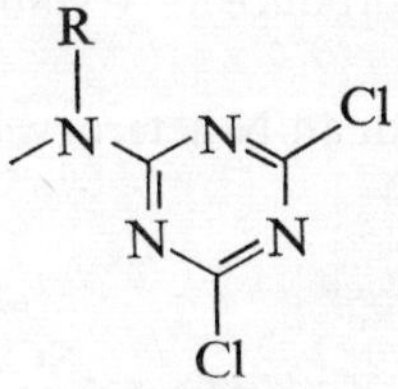

(4.27)

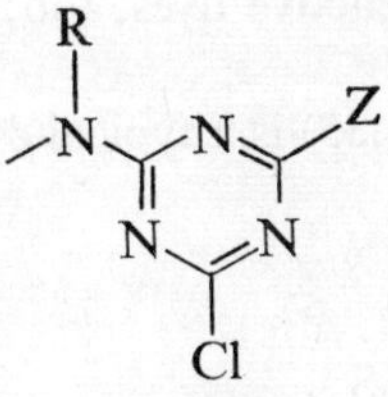

(4.28) Z: NR′R″ or OR‴

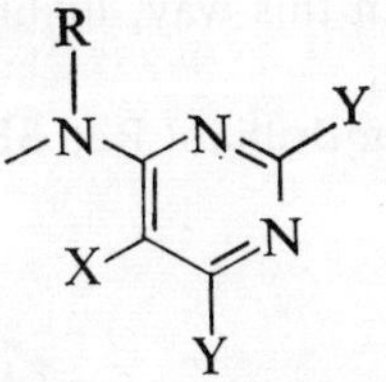

(4.29a) X = Y = Cl
(4.29b) X = Cl; Y = F

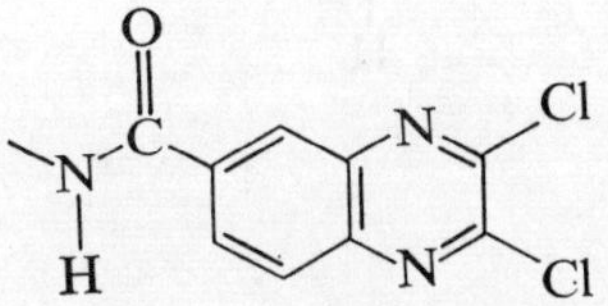

(4.30)

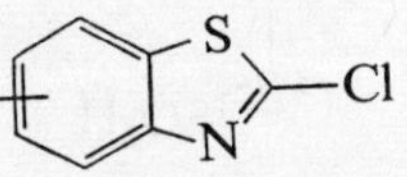

(4.31)

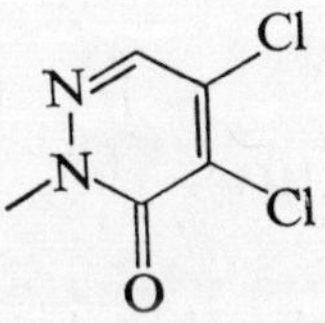

(4.32)

The reactivity of these N-heterocyclic reactive groups can be increased at the appropriate temperature—in cases where the addition reaction (k_1 in 4.11) has a limiting effect on the rate—by making the electrophilic C-atom more positive. This can be achieved by:

(a) additional electron-attracting cyclic nitrogen atoms (cf. conversion of pyrimidine into triazine groups),
(b) the introduction of electron-attracting substituents R†,
(c) the use of strongly electron-attracting leaving groups such as quaternary ammonium groups (e.g. **4.33** and **4.34**).

It is possible to obtain **4.33** or **4.34** by reacting the appropriate chlorinated N-heterocyclic groups with 1,4-diazabicyclo-[2,2,2]-octane ('DABCO'), or with asymmetrical dimethylhydrazine. Since the amine is again liberated in the reaction with the fibre, catalytic amounts suffice to accelerate fixation.[29]

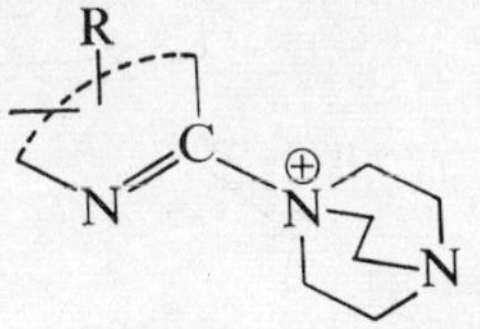

(4.33)

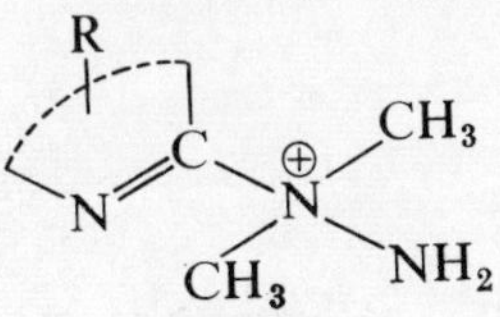

(4.34)

In the cases of (a) and (b), hydrolysis of the reactive dyeing (4.11: e.g. X = O-cellulose; Y = OH) is also accelerated, so that, generally speaking, the strength of

† Where the elimination step exerts a limiting effect on velocity (k_3 or also k_{-2} in (4.11)) the influence of substituents on reactivity may be reversible.

the covalent dye-fibre linkage[30] diminishes with increasing reactivity of dye. In the case of (c) this inverse proportionality between reactivity and stability can be avoided because the activating effect of the leaving group is lost during the fixation reaction. In this way, highly-reactive dyes, too, can produce very stable reactive bonds.

Cibacron Brilliant Red 3B (**4.35**) is a typical dye with an N-heterocyclic reactive group.

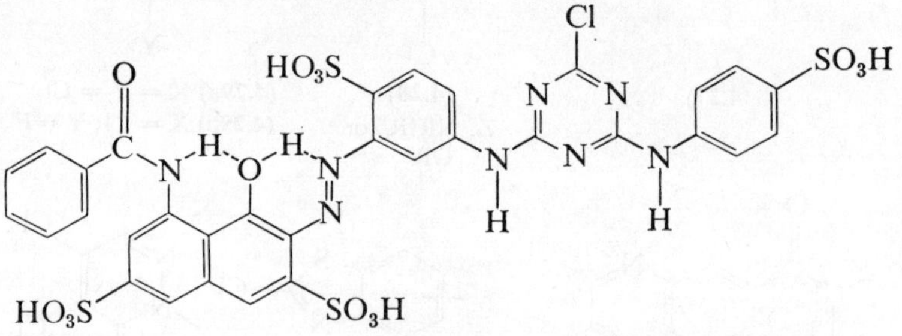

(4.35)

The principle underlying the synthesis of these reactive dyes is first to prepare the chromogenic fraction with at least one free amino group, which can then be reacted with, for example, cyanuric chloride (2,4,6-trichloro-*s*-triazine) to give the dichlorotriazine dye. Monochlorotriazine dyes can be prepared from the appropriate dichlorotriazine dyes by reacting them with a primary, secondary aliphatic or aromatic amine (or other nucleophilic agents). It is, however, also possible first to react cyanuric chloride with a colourless amine (or other nucleophilic agent) and to react the resulting dichlorotriazine derivatives in a second step, with the amino group of the dye. Dyes with other N-heterocyclic reactive groups are prepared similarly.

II. Groups that react in accordance with a nucleophilic addition mechanism, where there is frequently an elimination step before the addition step:
– the general base catalysed elimination of a nucleofugal leaving group (4.12: k_1)
– the specific base catalysed addition of the functional group of the textile fibre (4.12: k_2).

$$-Z-CH_2-CH_2-X \underset{k_{-1}}{\overset{k_1}{\rightleftarrows}} -Z-CH=CH_2 \begin{array}{c} \xrightarrow{+HY/k_2} \\ \xleftarrow{k_{-2}} \\ \xrightarrow{+H_2O/k_3} \end{array} \begin{array}{l} -Z-CH_2-CH_2-Y \quad (4.38) \\ \\ -Z-CH_2-CH_2-OH \quad (4.39) \end{array} \quad (4.12)$$

(4.36) (4.37)

Z: bridge link to the dyestuff.
HY: nucleophilic functional groups of the textile substrate

Azo Dyes

The systems **(4.40)**–**(4.45)** are important representatives of this reactive group.

—SO$_2$—CH$_2$—CH$_2$—OSO$_3$H   —NH—CO—CH$_2$—CH$_2$—OSO$_3$H
(**4.40**)   (**4.41**)
(Remazol dyes)   (Primazin dyes)

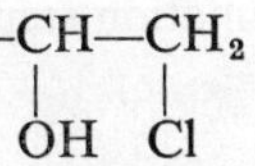

—CH—CH$_2$ —CH—CH$_2$
  | | \ /
OH   Cl   O
(**4.42**) 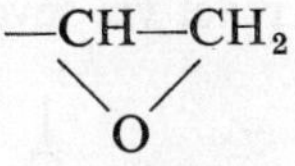   (**4.43**)

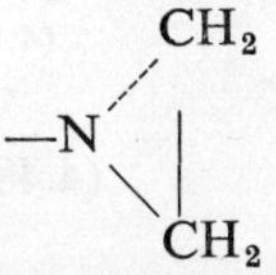

—NH—CH$_2$CH$_2$—Cl   —N(CH$_2$)(CH$_2$) (ring)
(**4.44**)   (**4.45**)
(Procinyl dyes)

An important difference between the fixation reactions (4.11) and (4.12) is that the functional group of the textile fibre in (4.11) participates in the addition as well as the elimination reaction, whilst in (4.12) the first elimination step is independent of the textile substrate. Therefore, with the reactive systems of group II there is, in principle, the possibility, at constant dyebath pH, of optimising relative to each other the rate of formation of the true reactive compound **4.37** and the rate of diffusion of the dye by varying the concentration and type of buffer; this may be of particular significance in obtaining level reactive dyeings. In contrast to (4.11), (4.12) shows, furthermore, that the degree of fixation must always depend on the type and concentration of buffer used in the dyeing process (general base catalysed reaction k_{-2} in (4.12)).

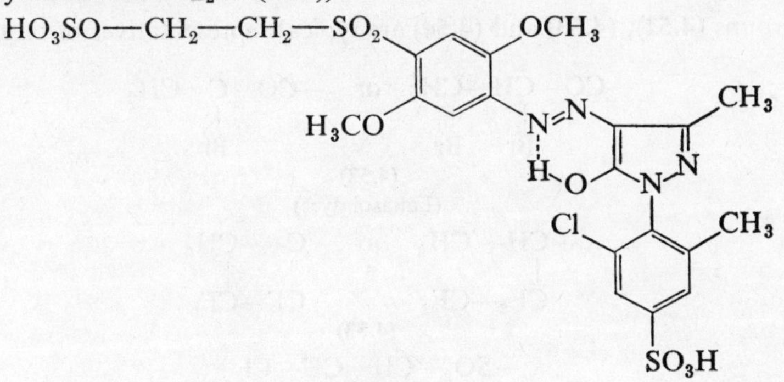

(**4.46**)

Remazol Golden Yellow G (**4.46**), a typical representative of this class of reactive dyes, is made by coupling diazotized 4-amino-2,5-dimethoxyphenyl-β-hydroxyethylsulphone with 1-(2′-chloro-4′-sulpho-6′-methyl)phenyl-3-methyl-5-pyrazolone.

The resulting dye is then esterified by dissolving it in concentrated sulphuric acid.

The β-hydroxyethylsulphone derivative is obtained by transforming an intermediate product into the sulphinic acid and reacting it with ethylene oxide, or by oxidizing β-hydroxyethyl sulphide, prepared by reacting a mercaptan with ethylene oxide.

III. Groups that react via several addition- and elimination steps (4.13):

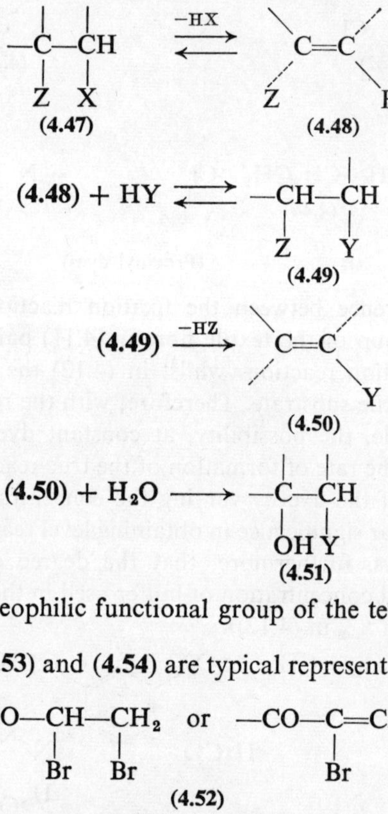

$$\underset{\underset{Z\ \ X}{|\ \ |}}{-C-CH} \underset{}{\overset{-HX}{\rightleftarrows}} \underset{Z\quad\quad H}{\diagdown C=C \diagup}$$
$$(4.47) \qquad\qquad (4.48)$$

$$(4.48) + HY \rightleftarrows \underset{\underset{Z\ \ Y}{|\ \ |}}{-CH-CH}$$
$$(4.49)$$

$$(4.49) \xrightarrow{-HZ} \underset{Y}{\diagdown C=C \diagup} \qquad\qquad (4.13)$$
$$(4.50)$$

$$(4.50) + H_2O \longrightarrow \underset{\underset{OH\ Y}{|\ \ |}}{-C-CH}$$
$$(4.51)$$

HY: nucleophilic functional group of the textile fibre

The groups (4.52), (4.53) and (4.54) are typical representatives of these systems.

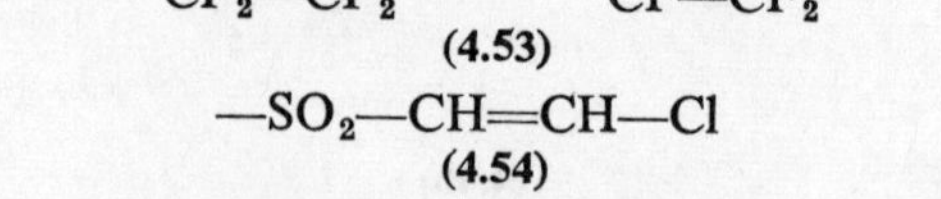

$$\underset{\underset{Br\ \ Br}{|\ \ |}}{-CO-CH-CH_2} \quad \text{or} \quad \underset{\underset{Br}{|}}{-CO-C=CH_2}$$
$$(4.52)$$
(Lanasol dyes)

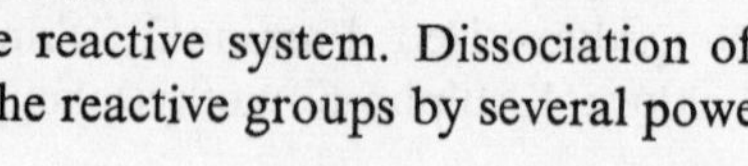

$$\underset{\underset{CF_2-CF_2}{|\ \ \ \ |}}{-CH-CH_2} \quad \text{or} \quad \underset{\underset{CF-CF_2}{|\ \ \ \ |}}{-C\ \ \ \ CH_2}\ (\|)$$
$$(4.53)$$

$$-SO_2-CH=CH-Cl$$
$$(4.54)$$

The bridge link: This influences:
(a) The reactivity of the reactive system. Dissociation of the imino bridge may reduce the reactivity of the reactive groups by several powers of ten. This, in many

IV. Groups that react with the fibre in acid conditions:

—NH—CH$_2$—OH
(4.55)

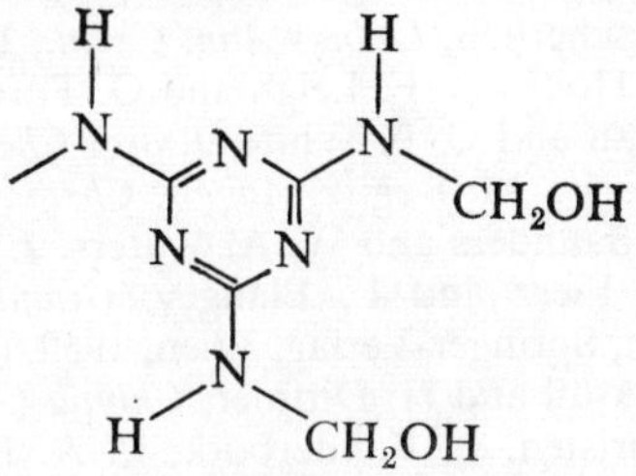

(4.56)
(Calcobound dyes)

V. Systems that, as polyfunctional fixing components, form covalent bonds with both the dyestuff residues and the substrate on the fibre.

$\boxed{F}$—SO$_2$NH$_2$ + CH$_2$=CH—CO—NR—R'—NR—CO—CH=CH$_2$ +

HO–cellulose

(4.57)
(Basazol dyes)

cases, also entails a diminution of substantivity, and can lead to lower fixation.[31,32,33]
(b) Selectivity (or degree of fixation). Model experiments have shown that N-heterocyclic reactive groups, that are linked via —NH— bridges to the dyestuff residue, are present in tautomeric forms of varying selectivity.[32] In this case, the reactive group has the greatest selectivity when the imino bridge is alkylated. On the other hand, alkylation generally reduces substantivity and hence degree of fixation. The length and flexibility of the bridge link also affect fixation.
(c) The stability of the reactive dyeing.[34] It is possible to split not only the dye-fibre bond but also the link between dye and reactive group.

Chromogenic (coloured) part of the molecule. This may belong to each of the classes listed in the chemical classification (cf. section 1.2) provided that it has the necessary affinity for the fibre. Generally speaking, the following are used: for yellow, orange and red, simple, metal-free monoazo dyes; for violet, ruby and navy, cupriferous mono- and disazo dyes; for bright blue, anthraquinone derivatives based on bromamine acid (section 10.54), also phthalocyanine derivatives (section 7.3). There are also reports on the influence of the chromophore on reactivity.[32] Dyestuff residues with a high association tendency, because of association preequilibria, reduce the rate of hydrolysis and frequently improve fixation owing to their higher substantivity. In some cases, however, undesirable dyestuff dimers may form.[35]

Water-solubilizing groups. With dyes for cellulose and protein fibres, 1 to 4 sulphonic groups are necessary, but with reactive disperse dyes for polyamide fibres special solubilizing groups are not generally required, since the reactive groups themselves confer sufficient solubility for the disperse dyeing process.

Summaries of information on problems relating to reactive dyes have been published.[33,36-38]

4.3. LITERATURE

1. E. Mitscherlich, *Liebigs Ann. Chem.*, **12**, 311 (1834).
2. H. H. Hodgson, F. Leigh and G. Turner, *J. chem. Soc. (London)* **1942**, 744; D. C. Freeman and C. E. White, *J. org. Chem.*, **21**, 379 (1956).
2a. Summary: M. Regitz, *Angew. Chem.*, **79**, 786 (1967).
3. K. H. Saunders and W. A. Waters, *J. chem. Soc. (London)*, **1946**, 1154.
4. H. E. Fierz and L. Blangey, *Grundlegende Operationen der Farbenchemie*, 8th edition, Springer-Verlag, Wien, 1952, (a) p. 328, (b) p. 243, (c) p. 244.
5. F. Suckfüll and H. Dittmer, *Chimia (Aarau, Schweiz)* **15**, 137 (1961).
6. M. Christen, L. Funderburk, E. A. Halevi, G. E. Lewis and H. Zollinger, *Helv. chim. Acta*, **49**, 1376 (1966).
7. T. Zincke and H. Bindewald, *Ber. dtsch. chem. Ges.*, **17**, 3026 (1884).
8. M. Martynoff, *Bull. Soc. chim. France*, **1951**, 214.
9. Summary: S. Hünig, H. Balli, K. H. Fritsch, H. Herrmann, G. Köbrich, H. Werner, E. Grigat, F. Müller, H. Nöther and K.-H. Oette, *Angew. Chem.*, **70**, 215 (1958); S. Hünig, H. Balli, E. Breither, F. Brühne, H. Geiger, E. Grigat, F. Müller and H. Quast, ibid., **74**, 818 (1962).
10. H. Balli and F. Kersting, *Liebigs Ann. Chem.*, **647**, 1 (1961); H. Balli, ibid., **647**, 11 (1961); H. Balli and F. Kersting, ibid., **663**, 96 (1963); H. Balli and R. Gipp, ibid., **699**, 133 (1966).
11. K. H. Saunders, *The Aromatic Diazo-Compounds*, 2nd edition, Edward Arnold & Co., London, 1949, p. 6.
12. H. Zollinger, *Chemie der Azofarbstoffe*, Birkhäuser-Verlag, Basel, 1958, (a) Chap. 2, (b) Chap. 4.3.
13. H. Zollinger, *Diazo and Azo Chemistry*, Interscience, New York, 1961, (a) Chap. 1, (b) Chap. 3.
14. E. D. Hughes, C. K. Ingold and J. H. Ridd, *J. chem. Soc. (London)*, **1958**, 58, 65, 70, 77, 82, 86.
15. J. H. Ridd, *J. Soc. Dyers Colourists*, **81**, 355 (1965).
16. R. Pütter, *Angew. Chem.*, **63**, 188 (1951); H. Zollinger, *Helv. chim. Acta*, **38**, 1597, 1623 (1955).
17. O. A. Stamm and H. Zollinger, *Helv. chim. Acta*, **40**, 1105, 1955 (1957).
18. H. Zollinger, *Chemie der Azofarbstoffe*, Birkhäuser-Verlag, Basel, 1958, Chap. 10; also *Chem. & Ind.*, **1965**, 885.
19. R. Kuhn and F. Bär, *Liebigs Ann. Chem.*, **516**, 143 (1935); A. Burawoy, A. G. Salem and A. R. Thompson, *J. chem. Soc. (London)*, **1952**, 4793, **1953**, 1443; D. Hadzi, *J. chem. Soc. (London)*, **1956**, 2143.
20. C. Müller, *Supplementum Chimia (Aarau, Schweiz)*, **1968**, 69.
21. Summary: A. Zitscher, *Melliand Textilber.*, **35**, 1389 (1954); W. Kirst and W. Neumann, *Angew. Chem.*, **66**, 429 (1954); M. Hueckel, *Textile Chem. Col.*, **1**, 510 (1969).
22. H. Zollinger, *Chimia (Aarau, Schweiz)*, **22**, 9 (1968).
23. H. Baumann and H. R. Hensel, *Fortschr. chem. Forsch.*, **7**, 643 (1967).
24. R. Grieb and A. Niggli, *Helv. chim. Acta* **48**, 317 (1965); H. Jaggi, *Helv. chim. Acta* **51**, 580 (1968); cf. also R. H. Prince and R. C. Spencer, *J. chem. Soc. (A)* **1968**, 2383.
24a. G. Schetty, *Helv. chim. Acta*, **53**, 1437 (1970).
25. G. Schetty, *Helv. chim. Acta*, **48**, 1042 (1965); **50**, 1039, 1836 (1967); *Textilveredlung* **3**, 3 (1968).

25a. R. Brändli, H. R. Oswald, R. Schweizer, P. Rys and H. Zollinger, *Helv. chim. Acta*, **53**, 1130 (1970).
26. W. Beckmann, *Melliand Textilber.*, **66**, 88 (1965).
26a. B. Gloor and H. Zollinger, *Helv. chim. Acta*, **54**, 553, 563 (1971).
27. O. A. Stamm, *Helv. chim. Acta*, **46**, 3008, 3019 (1963); G. Reinert, K. Mella, P. F. Rouette and H. Zahn, *Melliand Textilber.*, **49**, 1313 (1968); J. Shore, *J. Soc. Dyers Colourists*, **84**, 408, 413, 545 (1968); **85**, 14 (1969).
28. H. C. Brown and C. R. Smoot, *J. Am. chem. Soc.*, **78**, 6255 (1956).
29. T. L. Dawson, *J. Soc. Dyers Colourists*, **80**, 134 (1964).
30. R. Ch. Senn and H. Zollinger, *Helv. chim. Acta*, **46**, 781 (1963); R. Ch. Senn, O. A. Stamm and H. Zollinger, *Melliand Textilber.*, **44**, 261 (1963).
31. H. Ackermann and P. Dussy, *Melliand Textilber.*, **42**, 1167 (1961); *Helv. chim. Acta*, **45**, 1683 (1962); S. Horrobin, *J. chem. Soc.*, **1963**, 4130; J. R. Aspland and A. Johnson, *J. Soc. Dyers Colourists*, **81**, 425, 477 (1965).
32. P. Rys, *Textilveredlung*, **2**, 95 (1967); P. Rys and H. Zollinger, *Helv. chim. Acta*, **49**, 749, 761 (1966); A. Datyner, P. Rys and H. Zollinger, *Helv. chim. Acta*, **49**, 755 (1966).
33. E. Siegel, *Chimia (Aarau, Schweiz)*, **22**, 100 (1968).
34. J. Benz, *J. Soc. Dyers Colourists*, **77**, 734 (1961); O. Thumm and J. Benz, *Angew. Chem.*, **74**, 712 (1962).
35. P. Rys and O. A. Stamm, *Helv. chim. Acta*, **49**, 2287 (1966).
36. H. Zollinger, *Angew. Chem.*, **73**, 125 (1961); A. Zenhäusern and H. Zollinger in *Ullmann's Encyclopädie der techn. Chemie*, 3rd edition, Vol. 14, 615; O. A. Stamm, *J. Soc. Dyers Colourists*, **80**, 416 (1964).
37. G. E. Krichevski, *Aktivnije Krasitjeli*, Izdatjelstvo Legkaja industrija, Moskow, 1968.
38. W. F. Beech, *Fibre-Reactive Dyes*, Logos Press, London, 1969.
39. P. Rys, A. Schmitz and H. Zollinger, *Helv. chim. Acta*, **54**, 163 (1971).

5

Nitro and Nitroso Dyes

THIS group comprises all those dyes whose colour originates solely from the fact that an electron donor is conjugated with a nitro or nitroso group, in most cases via a system of aromatic π-electrons. One can distinguish the hydroxynitro (or nitroso) and aminonitro dyes according to the type of electron donor; these are mainly yellow-brown to brown acid, disperse and pigment dyes.

Whilst commercial hydroxynitro dyes are derivatives of 2,4-dinitro-1-naphthols, the most important aminonitro dyes are derived from 4-nitrodiphenylamine, e.g. Amido Yellow E (C.I. Acid Orange 3) (**5.1**).

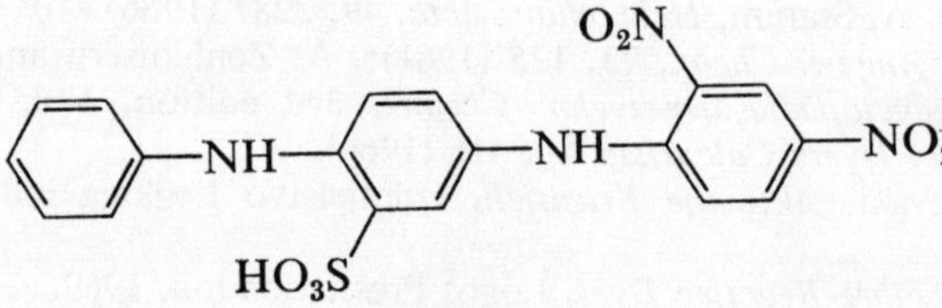

(5.1)

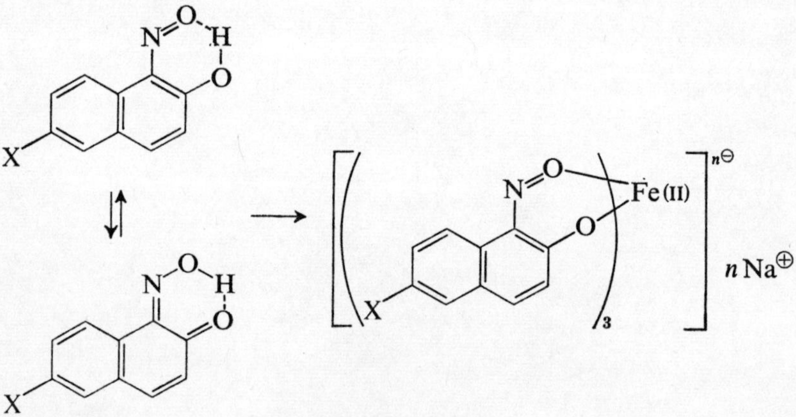

(5.2)
(a) X = H : $n = 1$
(b) X = SO$_3$H : $n = 4$

(5.1)

The hydroxy nitroso compounds, which, owing to their tautomeric equilibria (cf. 5.1) can also be considered to be quinonoximes, are, at present, used exclusively as metal complex dyes. The most important representatives of this series are Pigment Green B (C.I. Pigment Green 8) (**5.2a**) and Naphthol Green B (C.I. Acid Green 1) (**5.2b**). These are two of the few dyes which can be applied as 1:3-ferro- as well as 1:3-ferri-complexes (formation of complexes cf. section 3.6).

6
Polymethine Dyes

POLYMETHINE dyes comprise a large group of coloured compounds, which, in a simplified form, can be described by means of the mesomeric system **(6.1)**.

$$\left[X=\overset{|}{C}-\left(\overset{|}{C}=\overset{|}{C}\right)_{\!n}\!\!-Y \right]^{q} \longleftrightarrow \left[X-\overset{|}{C}=\left(\overset{|}{C}-\overset{|}{C}\right)_{\!n}\!\!=Y \right]^{q}$$

(6.1)

The groups X and Y are linked by a conjugated chain with an odd number of methine groups†; one acts as electron acceptor, the other as electron donor.

Although the arylcarbonium (or azenium) dyes can also be described with the aid of **(6.1)** ($n = 5$ with two benzenoid and quinonoid methine groups each) it is preferable to reserve their discussion to a separate chapter (see Chapter 8).

6.1. GENERAL DISCUSSION AND STRUCTURES

In accordance with the charge q in **(6.1)**, polymethine dyes can be sub-divided into cationic (q positive), anionic (q negative) or neutral ($q = 0$) compounds. The length of the methine chain also varies, giving rise to the vinyl homologues, mono- ($n = 0$), tri- (or carbo-; $n = 1$), penta- (or dicarbo-; $n = 2$) etc. methines. Many of these classes of dyes also possess trivial names, but only those in common use will be considered here. They can be further subdivided in accordance with the groups X and Y (cf. **(6.1)**). In the most important group of cationic polymethine dyes X and Y contain nitrogen. According to whether the nitrogen is a component of a ring, or not, they are called streptocyanines **(6.2)**, cyanines **(6.3)** or hemicyanines **(6.4)**. Compounds having the structure **(6.2)** are vinyl homologues of the amidinium ion (**(6.2)**: $n = 0$) and are used as intermediates in the synthesis of long-chain cyanine dyes (cf. **(6.18)**, section 6.2).

The nitrogen atoms in **(6.3)** belong either to identical (symmetrical cyanines) or to different (asymmetric cyanines) heterocyclic groups. The most important of these are the thiazole, thiodiazole, pyrrole, pyrroline, indole, 1,3,3-trimethylindoline, benzimidazo-thiazole, -oxazole, -selenazole, tetrazole, pyrimidine, pyridine, quinoline and higher fused-ring N-heterocyclic systems. The polymethine chain is generally linked in the α-position to the nitrogen of the heterocyclic group (as can be seen from **(6.3)** and **(6.4)**); however, β- and γ-derivatives are also known.

† With aza derivatives, individual (one or two) methine groups are replaced by aza groups.

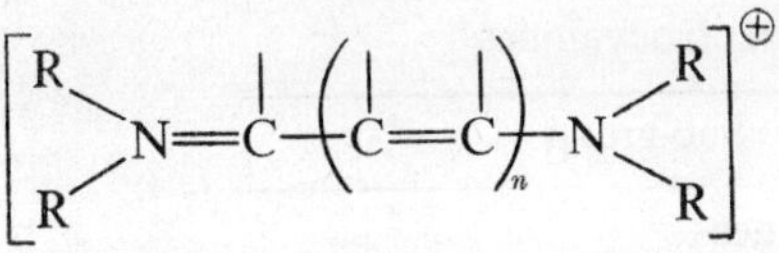

Polymethine Dyes

$$\left[\begin{array}{c} R \\ R \end{array} \!\!\!\! N=C\!\!\left(\!\!\begin{array}{c} | \\ C=C \\ | \end{array}\!\!\right)_{\!\!n}\!\!N\begin{array}{c} R \\ R \end{array} \right]^{\oplus}$$

(6.2)

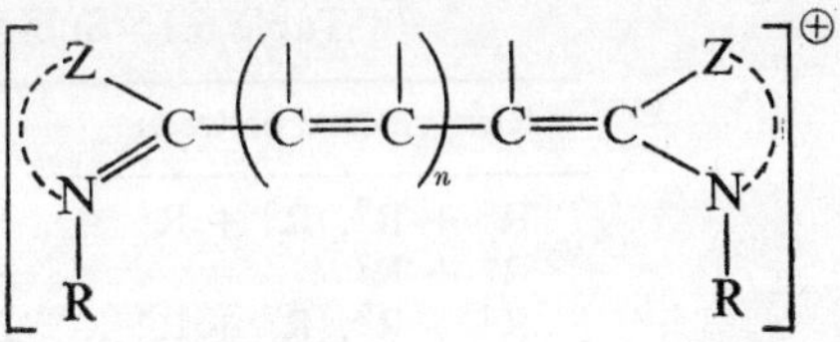

$$\left[\begin{array}{c} Z \\ N \\ | \\ R \end{array} \!\!\!\! C\!\!\left(\!\!\begin{array}{c} | \\ C=C \\ | \end{array}\!\!\right)_{\!\!n}\!\!C=C\begin{array}{c} Z \\ N \\ | \\ R \end{array} \right]^{\oplus}$$

(6.3)

$$\left[\begin{array}{c} Z \\ N \\ | \\ R \end{array} \!\!\!\! C\!\!\left(\!\!\begin{array}{c} | \\ C=C \\ | \end{array}\!\!\right)_{\!\!n+1}\!\!N\begin{array}{c} R \\ R \end{array} \right]^{\oplus}$$

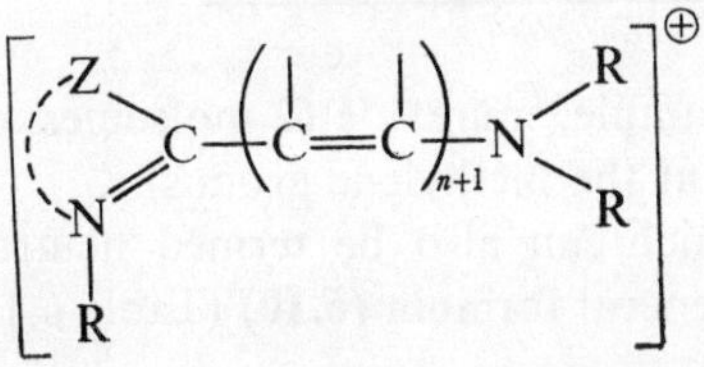

(6.4)

$Z = N, O, S, Se, C(CH_3)_2, CH=CH$

(Hemi-)cyanine-like polymethine dyes, in which X and Y (cf. **(6.1)**) contain oxygen, sulphur, phosphorus, or arsenic in place of nitrogen, have also been described.[1,7,9] Replacement of one or several methine groups with nitrogen in **(6.3)** or **(6.4)** gives rise to the so-called polyazacyanines or polyazahemicyanines.

The oxonols are the most important group of anionic polymethine dyes. They are vinyl homologues of the carboxylate ion (cf. **(6.5)**). In suitable oxonol dyes **(6.6)**, the last two carbon atoms at the ends of the polymethine chain are components of a heterocyclic group (cf. **(6.6)**).

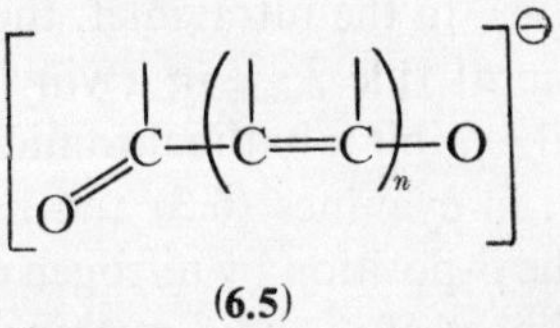

(6.5)

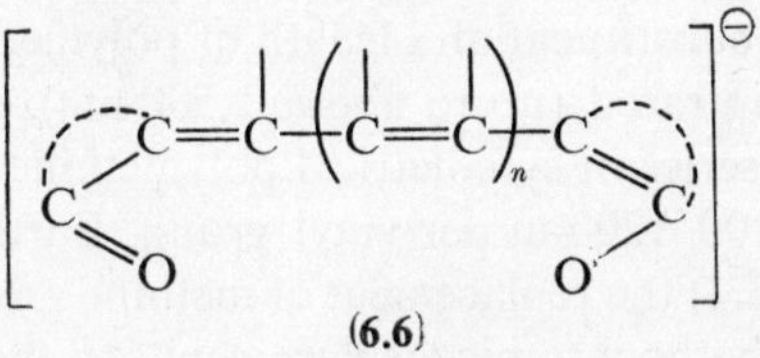

(6.6)

Some of the cyclic compounds which were used in the synthesis of **(6.6)** contain active methylene groups (e.g. **(6.7)**–**(6.9)**). The corresponding aza compounds of

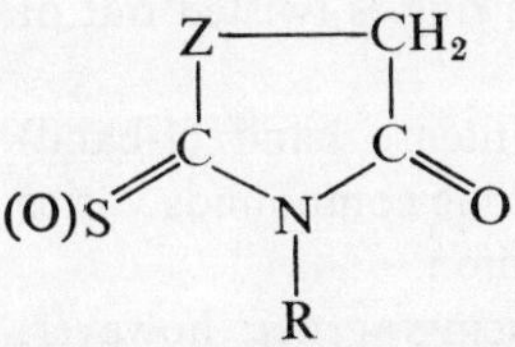

(6.7)
Z = O, S, Se, NR

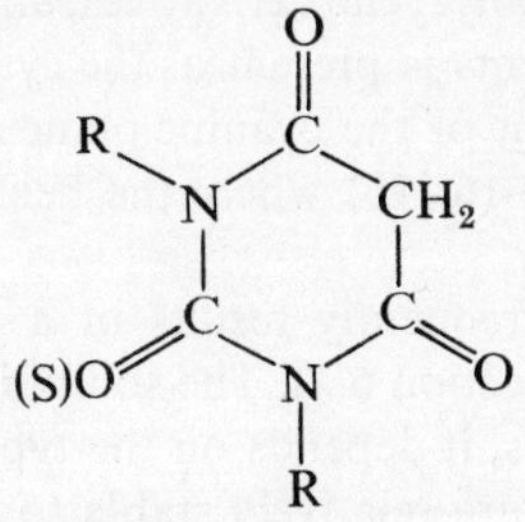

(6.8)

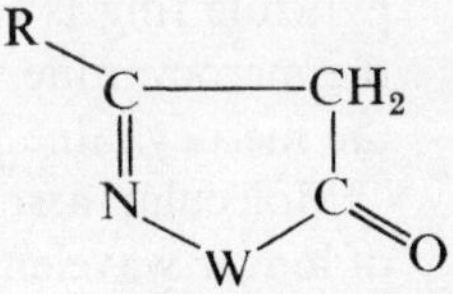

(6.9)
W = O, NR

Table 6.1 Sub-groups of neutrocyanines

Ring closure between	Sub-group
$R^2 + R^3$, $R^6 + R^7$	Merocyanines
$R^6 + R^7$	Hemioxonols
$R^2 + R^3$, $R^7 + R^8$	Dequaternated-cyanine-aldehydes
$R^2 + R^3$	(-Anil)

anionic polymethines are also known, as, for example, murexide (2 molecules of barbituric acid (**6.8**) are linked by an aza bridge at the methylene groups).

The neutral (non-ionic) polymethine dyes, which can also be termed neutrocyanines or merocyanines are described by the general formula (**6.10**) (Table 6.1).

$$R^1-Y\underset{n}{\underbrace{\left(\overset{(R^2)\ R^3}{\underset{|}{C}}=\overset{R^4}{\underset{|}{C}}\right)}}\underset{m}{\underbrace{\left(\overset{R^5}{\underset{|}{C}}=\overset{R^6}{\underset{|}{C}}\right)}}\overset{R^7\ (R^8)}{\underset{|}{C}}=Z$$

(**6.10**)

Y and Z = N, O, S, Se

The optical properties of individual polymethine dyes vary widely because of the multiplicity of structural differences. Their absorption spectra have been evaluated with the aid of the free electron model of H. Kuhn[2] (cf. section 2.1). The absorption band (λ_{max}) of the longest wavelength can lie, in accordance with the structure (e.g. type of substituent and length of polymethine chain), in the ultraviolet, the visible or the infrared (up to about 1,500 nm). As a general rule λ_{max} in a vinyl homologous series of symmetrical polymethine dyes will exhibit a bathochromic shift of about 100–120 nm per vinyl group. Furthermore, in cyanines (**6.3**) and hemicyanines (**6.4**) the replacement of methine groups in the β-position by nitrogen causes a strong bathochromic displacement and the replacement of α- and β-methine groups at the same time, as also of those in the α-position, gives rise to a hypsochromic shift. The absorption spectra of polymethine dyes generally show a positive solvatochromic effect. With polycyclic derivatives, an 'allopolar' isomerism may occur (**6.1**). (**6.11**) in polar solvents is present as the cyanine structure (**6.11a**) where the pyrazole ring is twisted out of the cyanine plane and in non-polar solvents it has the merocyanine structure (**6.11b**) where the benzothiazole ring is twisted out of the merocyanine plane.

Molecular association frequently results in a narrow, intense band (J-band) of longer wavelength (cf. section 6.4). The solubility of methine compounds varies widely and, with ionic dyes, it depends on the type of gegenion.

Most dyes of this class are relatively stable to mild reducing agents; however, acids will decolourise them reversibly and alkalis and oxidants irreversibly. In general, resistance diminishes with increasing number of methine groups.

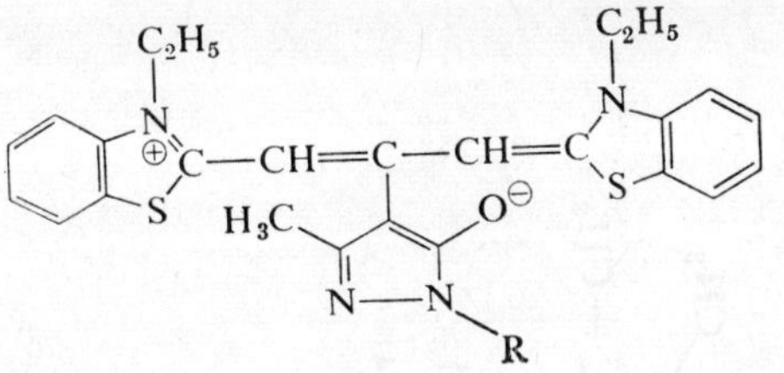

Polymethine Dyes

(6.11a) 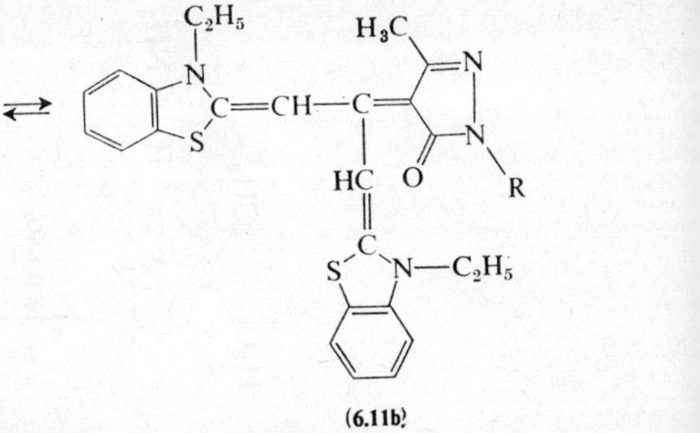 ⇌ (6.11b)

(6.1)

Numerous well-known natural dyes with a hemi-oxonol structure are derivatives of γ-pyran. Quercetin **(6.12)** and cyanidine **(6.13)** are so-called pyrylium compounds.

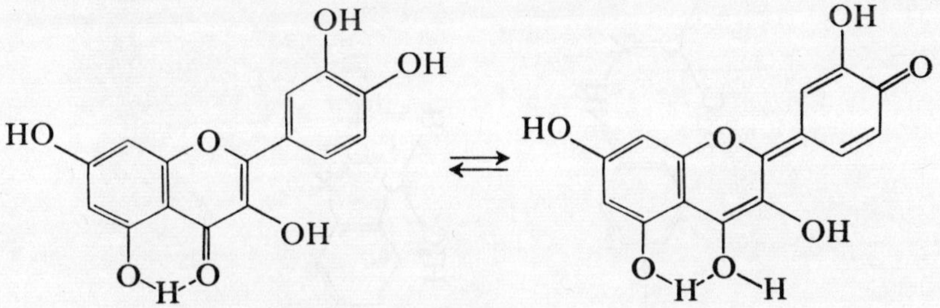

(6.12)

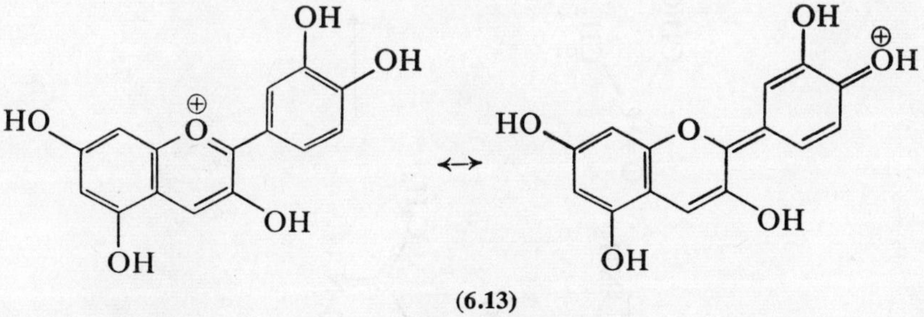

(6.13)

The history of polymethine dyes is characterized by four phases of development. After G. Williams, in 1856, discovered the first cyanine, in which two quinoline rings are linked by a methine group in the 4,4′-position, H. W. Vogel discovered, in 1873, that this and similar dyes possess a photosensitizing action (cf. section 6.4). The final elucidation of the constitution of cyanines by A. Kaufmann and E.

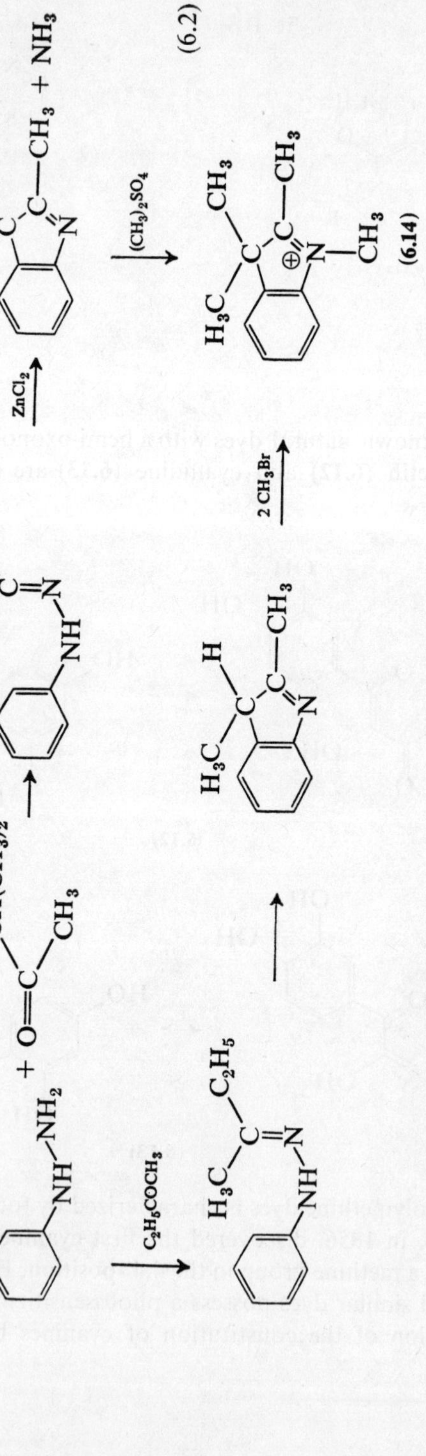

78   *Fundamentals of the Chemistry and Application of Dyes*

$$\text{PhNH-NH}_2 + \text{O=C(CH}_3)\text{CH(CH}_3)_2 \longrightarrow \text{H}_3\text{C-C(=N-NHPh)-CH(CH}_3)_2 \xrightarrow{\text{ZnCl}_2}$$

$$\begin{array}{c}\text{H}_3\text{C}\diagdown\text{C}\diagup\text{CH}_3\\ \diagup\text{C}=\text{N}\diagdown\\ \text{(benzopyrroleninium)}\end{array} + \text{NH}_3 \quad (6.2)$$

$$\xrightarrow{(\text{CH}_3)_2\text{SO}_4} \quad \begin{array}{c}\text{CH}_3\\ \text{H}_3\text{C}-\text{C}-\text{CH}_3\\ \text{C}=\overset{+}{\text{N}}-\text{CH}_3\end{array} \quad (6.14)$$

$$\text{PhNH-NH}_2 + \text{C}_2\text{H}_5\text{COCH}_3 \longrightarrow \text{H}_3\text{C-C(=N-NHPh)-C}_2\text{H}_5 \xrightarrow{2\,\text{CH}_3\text{Br}} \begin{array}{c}\text{H}\\ \text{H}_3\text{C}-\text{C}-\text{CH}_3\\ \text{C}=\text{N}\end{array}$$

Vonderwahl, in 1912, showed the way to the planned synthesis of polymethine dyes. The last period of development was begun in the 1920s by W. König (using o-formic acid esters) and has yielded a great number of polymethine derivatives.

6.2. THE PRINCIPLES UNDERLYING THE PREPARATION OF POLYMETHINE DYES

Because there is a very wide choice of heterocyclic groups, length of polymethine chains and type of substituents in the synthesis of polymethine dyes, it is quite impossible to give an exhaustive description of all synthetic methods. Those that are discussed now are, therefore, generalizations and no details of the precise reaction conditions will be given.

Equations (6.5) and (6.6) show that the heterocyclic compounds in the different syntheses are used as nucleophilic (cf. (6.5)) and electrophilic (cf. (6.6)) reagents. The 1,3,3-trimethyl-2-methylene indoline ((**6.15**): Fischer base, synthesis cf. (6.2)) is a heterocyclic system typical for these syntheses. The true condensation reactions are preceded by acid-base as well as dimerization equilibria as can be seen from equation (6.3) and the example (**6.15**).

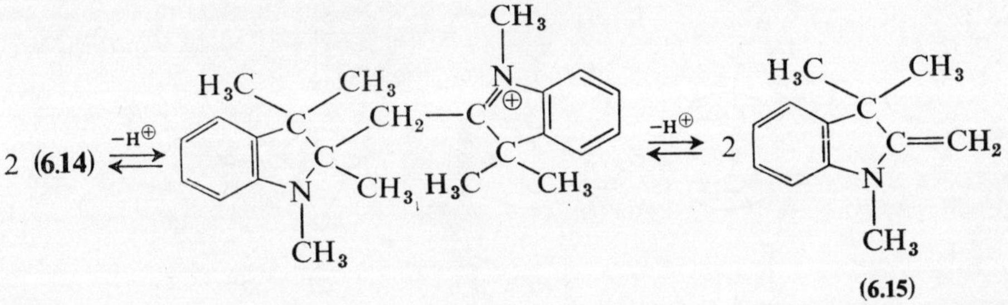

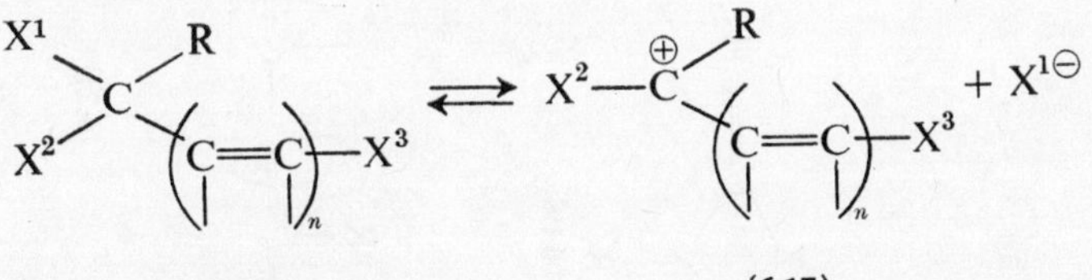

The compounds used in the synthesis of a polymethine chain (**6.17**) (or (**6.16**)) and (**6.18**) are, for example, chloroform (**6.16**: $R = H, X^1 = X^2 = X^3 = Cl, n = 0$), o-formic acid ester (**6.16**: $R = H$, $X^1 = X^2 = X^3 = OC_2H_5$, $n = 0$), vinylhomologue acid chlorides (**6.18**: $Y = 0$, $X^3 = Cl$, $n = 0, 1, 2 \ldots$), streptocyanine (**6.18** = **6.2**) or chloramides of the dialkyl (or diaryl) formamides and their homologues (**6.18**: $Y = NR^1R^2$, $X^3 = Cl$, $n = 0, 1, 2 \ldots$).

As equations (6.2)–(6.6) for cyanine show, oxonols (6.6) and merocyanines (6.10) are also synthesized. The various syntheses of merocyanines of a higher order

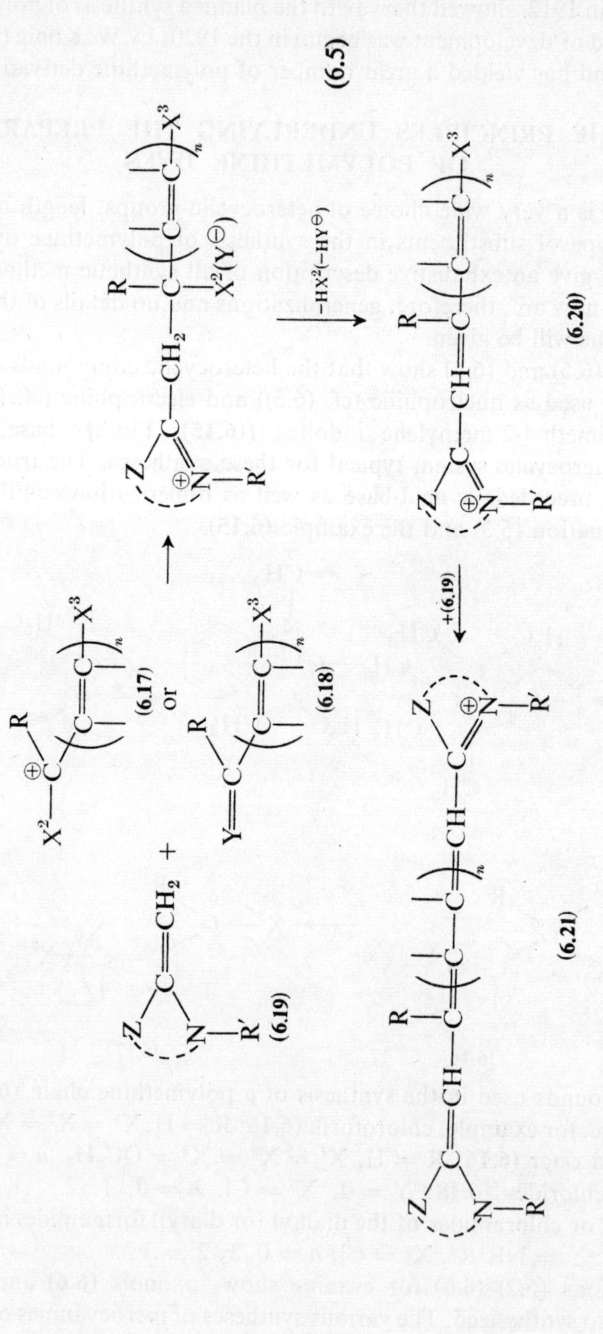

80 *Fundamentals of the Chemistry and Application of Dyes*

Polymethine Dyes

are interesting ((**6.26**), cf. (6.7)–(6.9)). Merocyanine (**6.24**) can be repeatedly reacted with Rhodanine ((6.7): $Z = S$, $R = C_2H_5$) which results in each case in a bathochromic shift of the absorption band by 20–50 nm. There are various comprehensive reviews of interesting aspects of the synthesis of polyazamethines.[1,3,7,9]

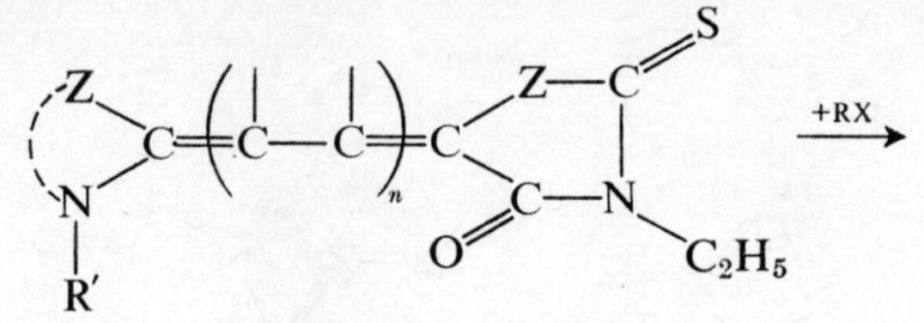

(6.24) $\xrightarrow{+RX}$

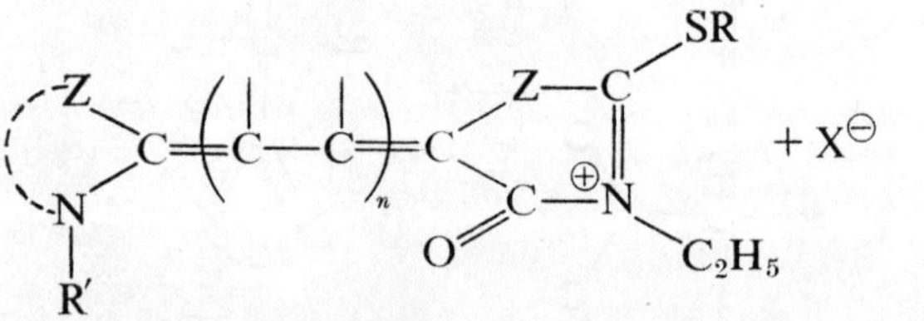

(6.25) $+ X^{\ominus}$ (6.7)

(6.25)
+
(6.7): $Z = S, R = C_2H_5$ $\longrightarrow$ (6.26)† (6.8)

(6.26) $\xrightarrow[\text{(cf. 6.7)}]{+RX}$ $\xrightarrow{+(6.19)}$

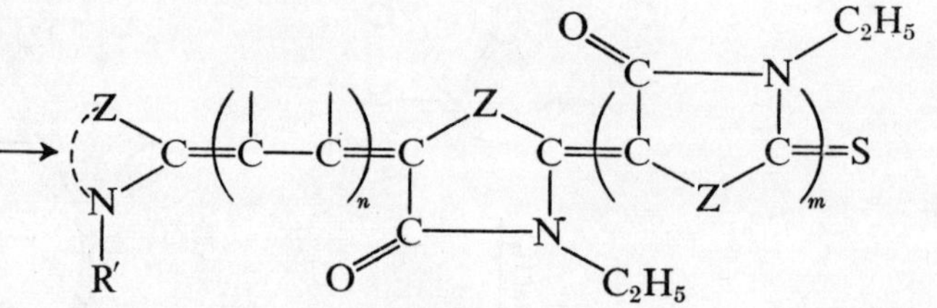

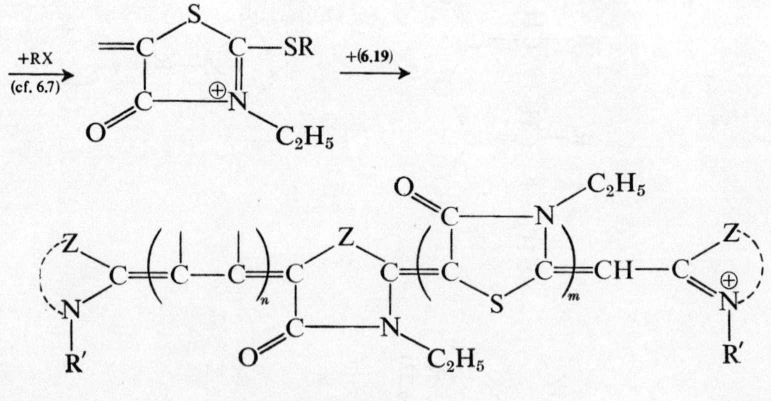

(6.27) (6.9)

† By repeating (6.7) and (6.8) several times: $m > 1$.

6.3. APPLICATION OF POLYMETHINE DYES IN DYEING

The application of polymethine dyes for the dyeing of textiles is limited by the fact that they generally have very poor light fastness. On the other hand, some of their aza analogues† have outstanding fastness. Individual representatives, e.g. (6.28) (λ_{max} in CH_3OH: 507 nm), have excellent light fastness with polyacrylonitrile fibres even in pale shades; Cyanine (6.29) (Astraphloxine FF—C.I. Basic Red 12) has unusual brilliance (it is a fluorescent dye) but poorer light fastness; the same

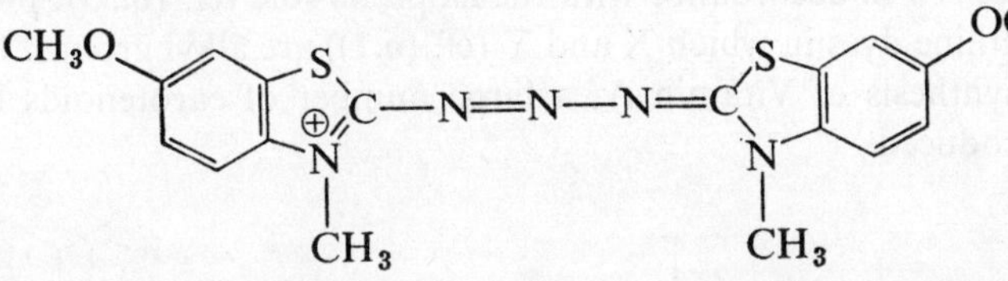

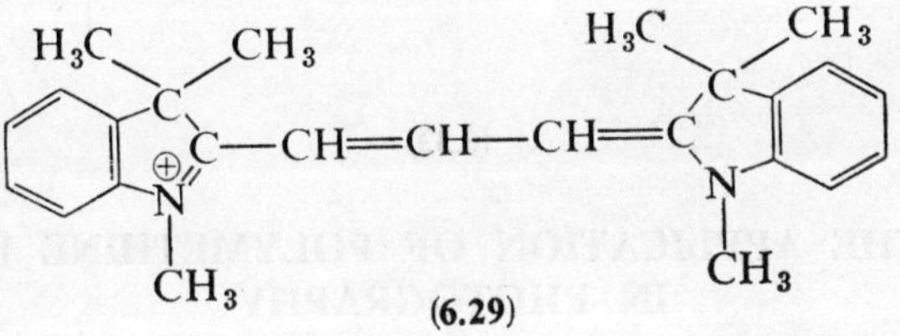

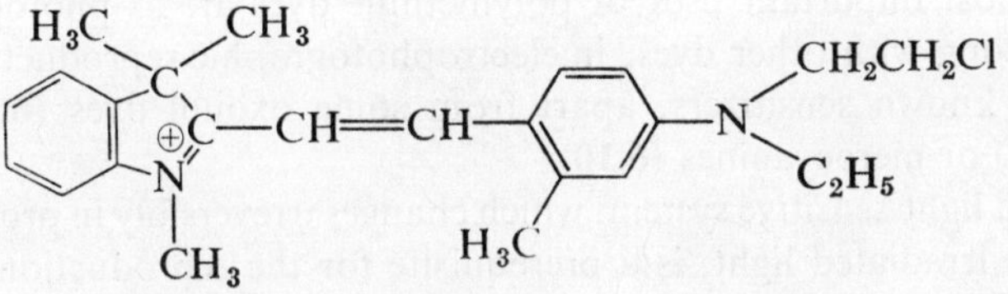

† The hydroxyazo dyes (cf. Chapter 4) can also be classified as monoazahemioxonols (cf. (6.10)). Similarly, aminoazo compounds can be formally treated as azacyanines. The practical significance of such considerations can be seen in the methods of synthesis of Maxilon dyes (cationic azo dyes). They can be produced either by quaternization (alkylation) of aminoazo compounds or by oxidative coupling by the method of Hünig (section 4.1.1).

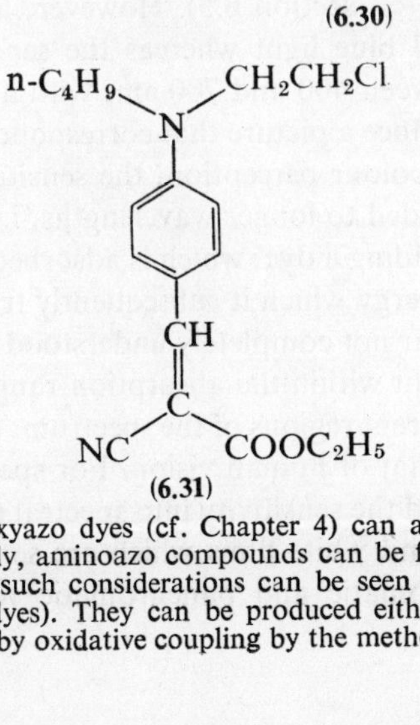

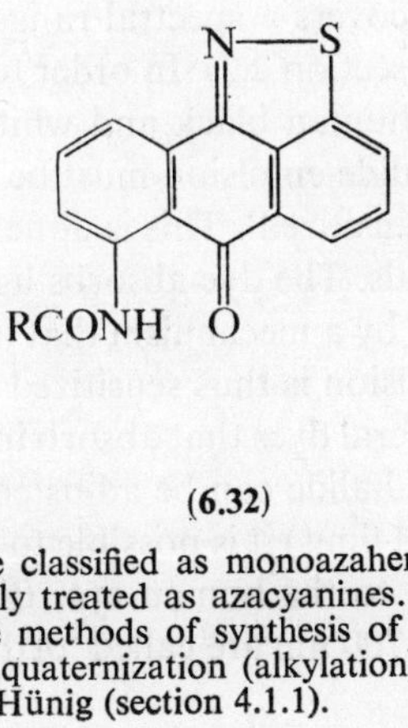

applies to the hemicyanine **(6.30)** (Astrazon Red 6B—C.I. Basic Violet 7), a polyacrylonitrile dye.

The disperse dyes **(6.31)** (Celliton Fast Yellow 7G 'F'—C.I. Disperse Yellow 31, a merocyanine) and **(6.32)** (a derivative of isothiazolylanthrone, an inner merocyanine) have adequate light fastness. Hydroxyazomethines are also classified as hemioxonols (cf. **(6.10)**); they have, however, no technical importance.

A group of natural dyes (the so-called carotenoids) are used for dyeing food. They are synthesized in accordance with the isoprene rule (cf. **(6.33)**: β-carotene), and are polymethine dyes in which X and Y (cf. **(6.1)**) are alkyl groups. In connection with the synthesis of Vitamin A, a large number of carotenoids have been successfully[4] produced.

(6.33)

6.4. THE APPLICATION OF POLYMETHINE DYES IN PHOTOGRAPHY

By far the most important uses of polymethine dyes are as photographic sensitizers and, along with other dyes, in electrophotographic reproduction processes. Most of the known sensitizers, apart from some oxonol dyes **(6.6)**, are either cyanines **(6.3)** or merocyanines **(6.10)**.

The use of a light-sensitive system, which changes irreversibly in proportion to the energy of the irradiated light, is a prerequisite for the reproduction of a picture. In photography a silver halide is used (cf. section 6.5). However, a silver halide emulsion reacts only to ultraviolet and blue light whereas the sensitivity of the human eye covers a spectral range between 400 and 700 nm, with a maximum at 555 nm (cf. section 2.2). In order to produce a picture that corresponds as nearly as possible to human black and white or colour perception, the sensitivity range of the silver halide emulsion must be extended to longer wavelengths, i.e. it has to be 'spectrally sensitized'. This is done by adding a dye, which is adsorbed on the silver halide crystals. The dye absorbs light energy which it subsequently transfers to the silver halide by a mechanism that is so far not completely understood (cf. reference 5). The emulsion is thus sensitized to light within the absorption range of the dye. By using several dyes that absorb in different regions of the spectrum, the sensitivity of the silver halide can be adjusted to that of human vision. For special purposes (e.g. infrared films) it is possible to extend the sensitivity into spectral regions which are invisible to the human eye. Black and white films which are sensitized up to 600 nm and 700 nm are called 'orthochromatic' and 'panchromatic' respectively.

Sensitization is frequently accompanied by side effects that are attributable, in part, to the steric arrangement of the adsorbed dyes. With suitable substitution of the polymethine chain, it is possible to enforce a planar structure, which is favourable to sensitization and adsorption and also to molecular aggregation. In many cases this causes a layer-like adsorption on the surface (like a pile of coins). Aggregation is indicated by the presence of a sharp absorption band (J-band), which is bathochromically displaced and depends on the concentration. The sensitizing action can be markedly increased (over-sensitization) or considerably reduced (anti-sensitization) by small amounts of foreign matter. The foreign material may transfer absorbed light energy to dye molecules that have not been exposed to light; therefore, the absorption of light and the sensitization can occur at different places.

Light impinging on photographic material can produce undesirable secondary latent pictures in places which should remain unexposed; this is due to scattering at the silver halide grains or to reflection at the film backing. In order to eliminate these 'halation' effects, which would cause blurring of the image, 'screening dyes' are added to the film material. These must be photographically inactive, exhibit the correct absorption spectrum and become completely decolourized when the film is developed. Apart from some triphenylmethane (cf. Chapter 8) and formazan dyes, representatives of the oxonols (**6.6**) and merocyanines (**6.10**) are particularly suitable for this purpose.

6.5. THE USE OF POLYMETHINE DYES IN COLOUR PHOTOGRAPHY

Dyes are used not only for sensitizing photographic materials (cf. section 6.4) but also for producing the image. In the most important methods of colour photography, the pictures are obtained, either by synthesis or bleaching of dyes in different light-sensitive layers, or by the transfer of dyes on to a receiving layer to form the picture.

Colour photography is based on the capacity of the human eye to perceive all visible colours occurring in nature by the addition of only three basic colours (blue, green and red), in different proportions (cf. section 2.2). It follows that a coloured picture represents a satisfactory copy of an object if, when it is irradiated with white light, it reflects or transmits the blue, green and red rays in such proportions that, as far as possible, they reproduce the colours of the original. The principal task of colour photography, therefore, consists of separating all the colours of the object being photographed into these three basic colours, to record their intensity separately, and to combine the resulting coloured sub-pictures to form the whole picture. This separation is achieved by photographing the object either successively or simultaneously by means of three different silver halide layers sensitized, respectively for blue, green and red light. Numerous processes are known for recording the sub-pictures.[6,8b,9] The most important operate on the principle of subtractive colour mixture: thus, for example, an amount of a yellow

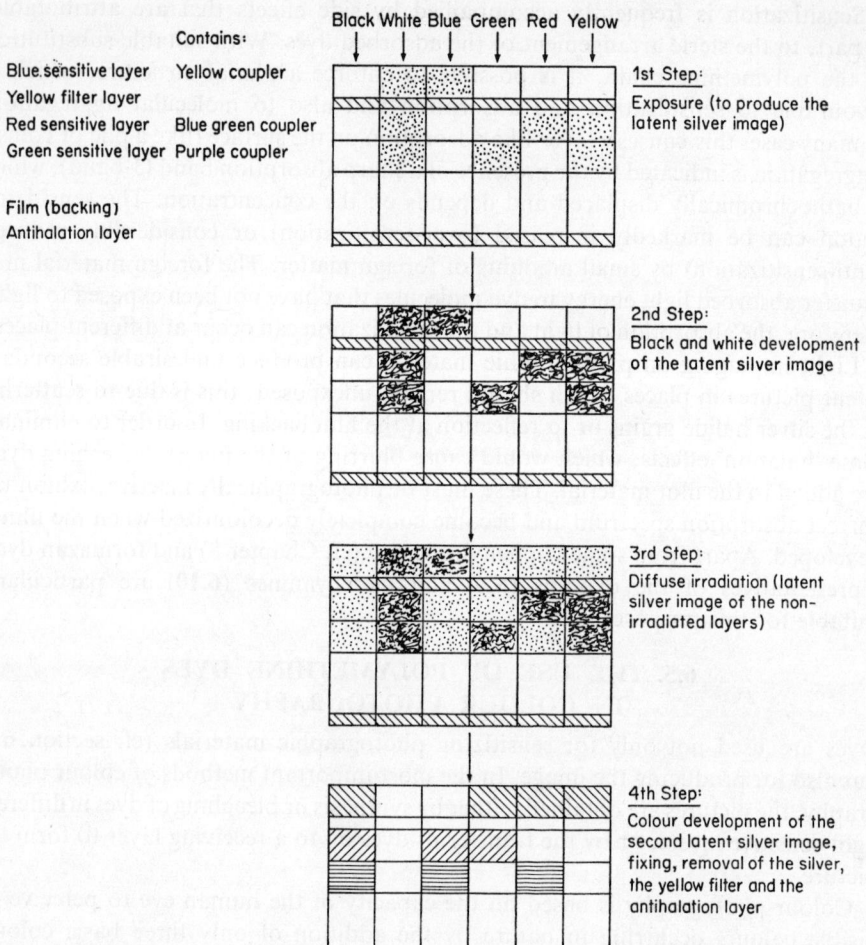

Figure 6.1 Exposure and development of a coloured film (reproduction of the positive).

dye is formed in the blue sensitized layer which is inversely proportional to the intensity of the blue light component; analogously purple and blue-green dyes, respectively, are formed in the green and red sensitized layers. The colour of the dye is complementary to that of the light for which the layer is sensitized. One of the numerous procedures for the production of the positive is based on a principle discovered in 1912 by R. Fischer (Figure 6.1).

The exposed material is first subjected to a black and white development in which the irradiated grains are reduced to metallic silver. If the coupling components for the subsequent dye synthesis are already present in the emulsion, this step requires a developer, the oxidation product of which will not react with them. The film is exposed to diffused light and colour development proper of the latent positive image follows mainly with N,N-dialkyl-*p*-phenylenediamine **(6.34)**, with heterocyclic hydrazones **(6.35)** or with tetracyanoethane **(6.36)**. Finally, the metallic silver is reoxidized to silver ions with the aid of potassium hexacyanoferrate (III) and eliminated from the emulsion by complex formation with sodium thiosulphate. The dye synthesis (6.11) conforms to the laws of electrophilic substitution (cf.

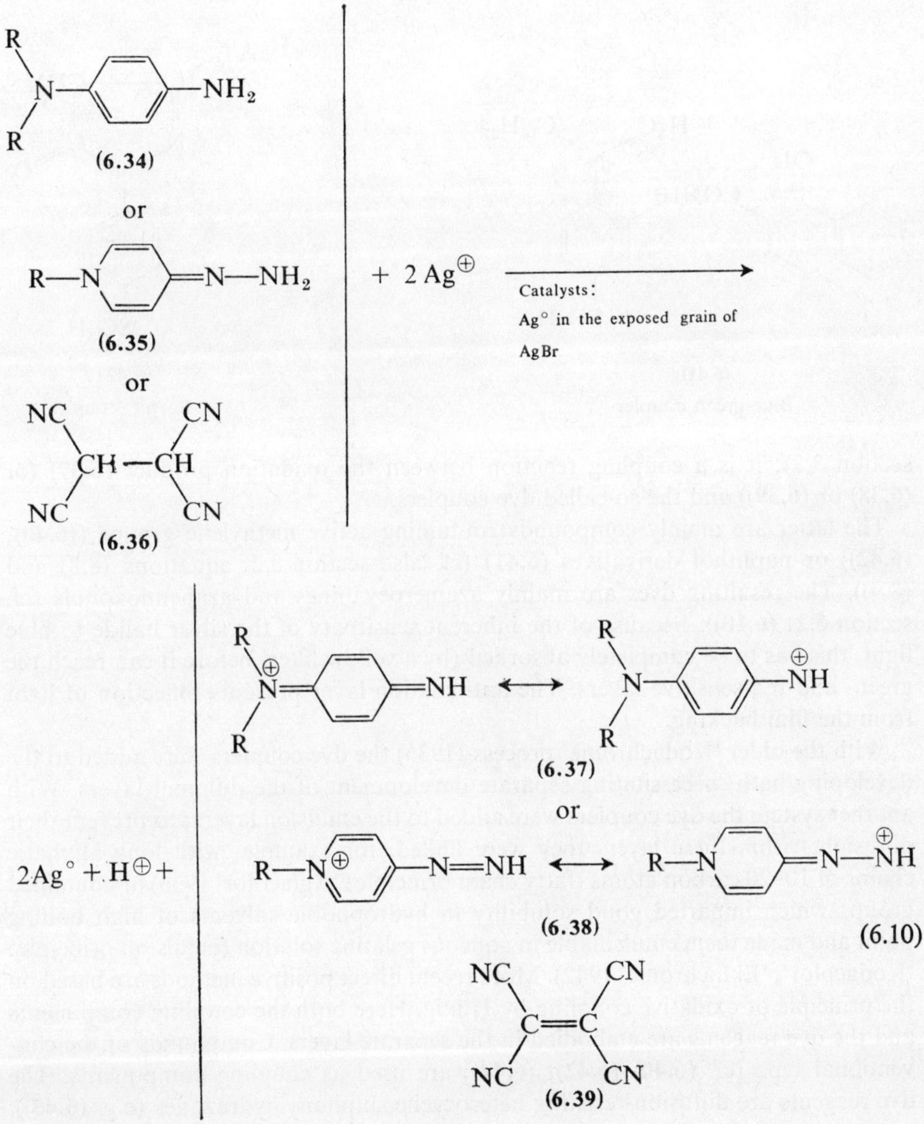

$$2\,Ag^\circ + H^\oplus + \text{[products (6.37), (6.38), (6.39)]} \quad \text{(6.10)}$$

Catalysts: Ag° in the exposed grain of AgBr

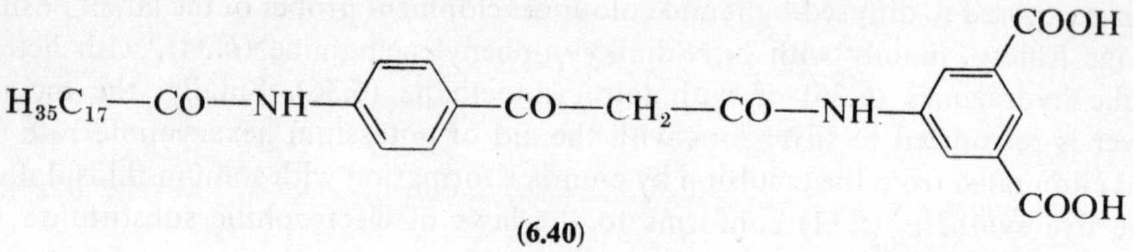

$H_{35}C_{17}$—CO—NH—⟨C₆H₄⟩—CO—CH$_2$—CO—NH—⟨C₆H₃⟩(COOH)(COOH)

(6.40)
Yellow coupler

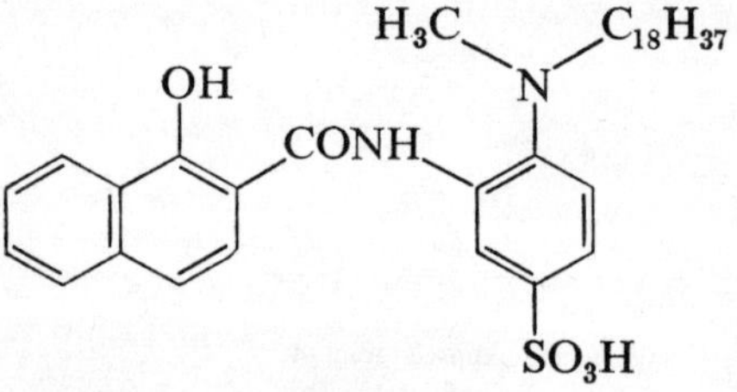

(6.41) Blue-green coupler: naphthol with OH and CONH linked to a phenyl ring bearing N(CH$_3$)(C$_{18}$H$_{37}$) and SO$_3$H.

(6.42) Purple coupler: pyrazolone with $H_{35}C_{17}$—C=N—N(Ar)—C(=O)—CHX, where Ar = C$_6$H$_4$—SO$_3$H.

(6.41)
Blue-green coupler

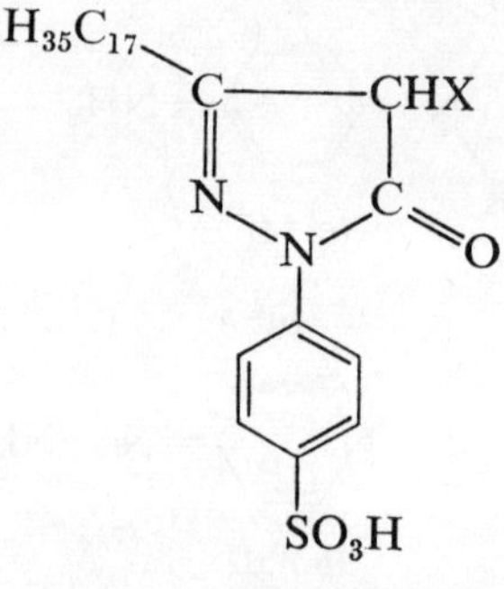

(6.42)
Purple coupler

section 3.3); it is a coupling reaction between the oxidation product **(6.37)** (or **(6.38)** or **(6.39)**) and the so-called dye couplers.

The latter are mainly compounds containing active methylene groups ((**6.40**), (**6.42**)) or naphthol derivatives (**6.41**) (cf. also section 8.2: equations (8.8) and (8.9)). The resulting dyes are mainly azamerocyanines and azahemioxonols (cf. section 6.2: (**6.10**)). Because of the inherent sensitivity of the silver halide to blue light, this has to be completely absorbed (by a yellow filter) before it can reach the green- and red-sensitive layers. The antihalation layer prevents reflection of light from the film backing.

With the older 'Kodachrome' process (1935) the dye couplers were added to the developing bath, necessitating separate development of the different layers. With another system the dye couplers were added to the emulsion layers; to prevent their diffusion within these layers they were linked, for example, with long aliphatic chains of 10–20 carbon atoms (fatty chain principle: 'Agfacolor' 1936) or contained groups which imparted good solubility in hydrophobic solvents of high boiling point and made them emulsifiable in aqueous gelatine solution (emulsion principle: 'Kodacolor', 'Ektachrome' 1942). More recent direct positive methods are based on the principle of oxidative coupling by Hünig. Here both the coupling components and the dye reagents are embodied in the separate layers. Compounds of the conventional type (cf. (**6.40**)–(**6.42**), (**6.44**)) are used as coupling components. The dye reagents are diffusion-resisting heterocyclic sulphonylhydrazones (e.g. (**6.45**)).

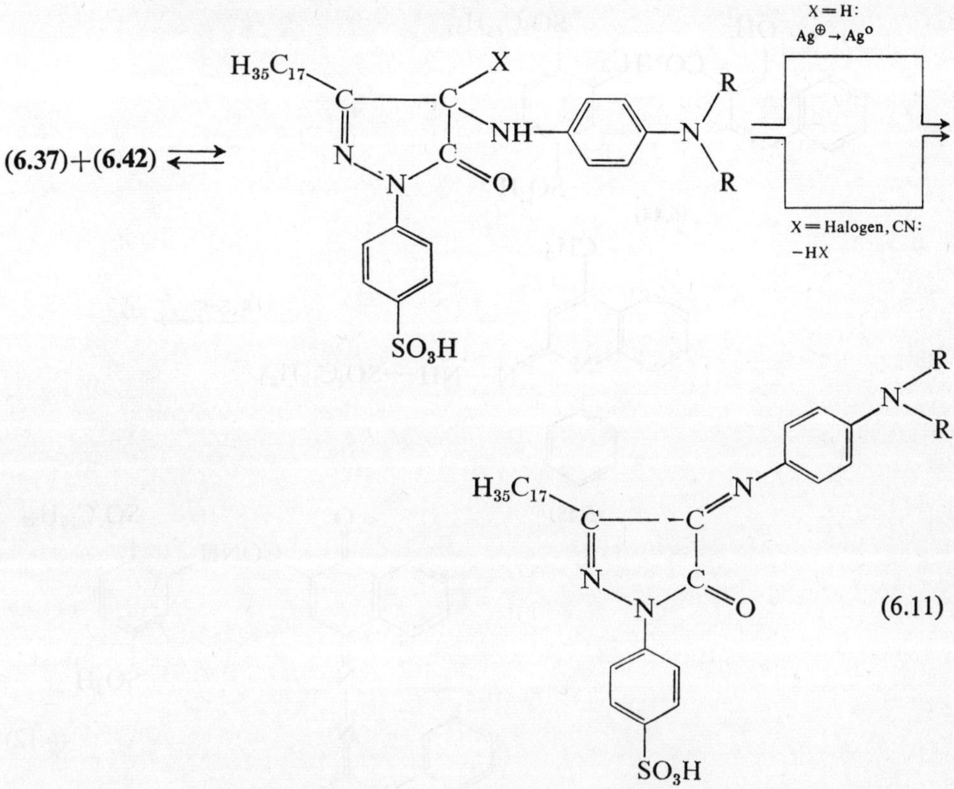

(6.37)+(6.42) ⇌

$X = H$: $Ag^{\oplus} \to Ag^{o}$

$X =$ Halogen, CN: $-HX$

(6.11)

(6.43)

In a first step the exposed material is developed with N,N-dialkyl-*p*-phenylene-diamine (**6.34**). Here, the quinone diamine (**6.37**), appropriate to the method that is being used, reacts with the coupling agent or the dye reagent. The amount of coupling agent or dye reagent consumed depends on the amount of light received by the film. The metallic silver is reoxidized in a solution of potassium hexacyano-ferrate (III) and the residual couplers and dye reagents are caused to react (6.12).

The compounds formed in the first step of development are then hydrolysed and the silver ions are removed by a fixing bath containing thiosulphate.

In some other photographic processes the coloured picture is produced by a light-catalysed dye bleaching process based on the fact that numerous dyes (e.g. azo dyes) are irreversibly destroyed by reduction with suitable baths in the presence of the silver which forms during exposure (e.g. dilute HBr baths and azines as hydride ion transfer catalysts).[10]

Coloured pictures can also be obtained by the diffusion of dyes into a receiving layer. With the 'Polaroid' process,[11] for example, the positive picture is formed by causing a reagent to penetrate into the three sensitized silver halide layers after

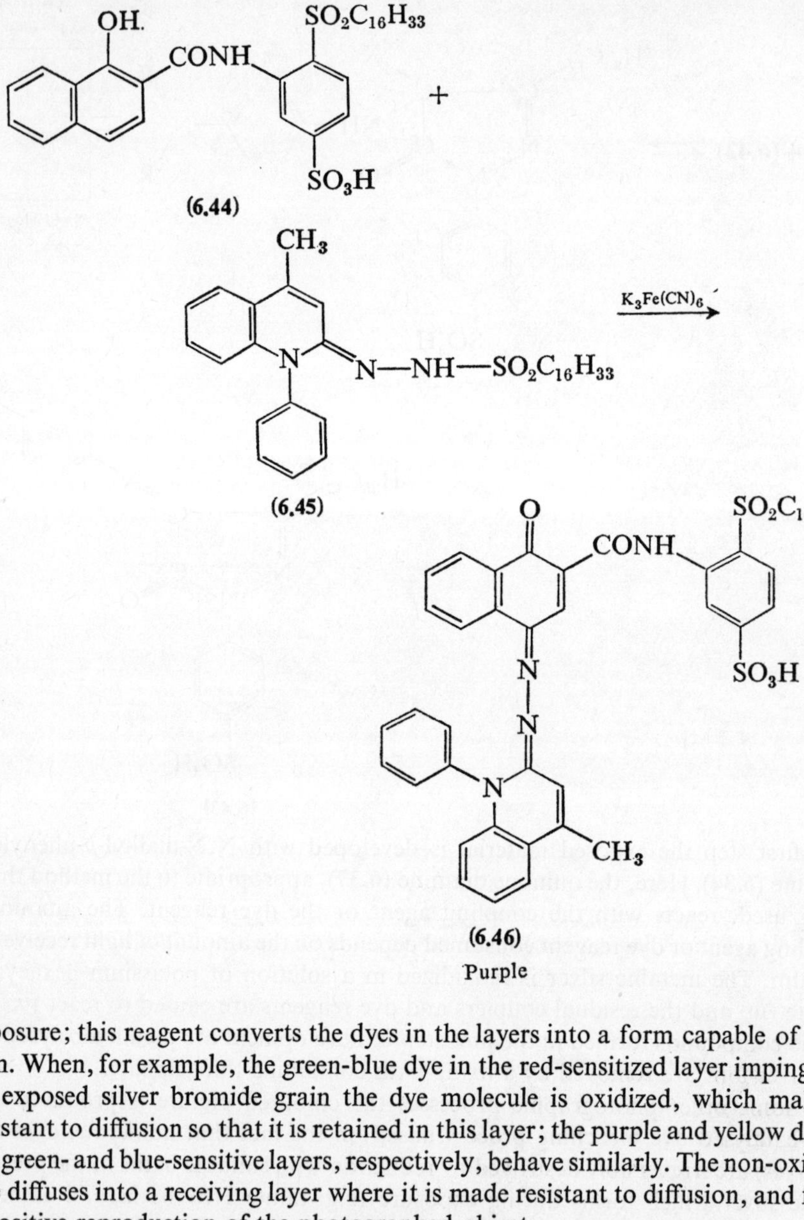

(6.44) + (6.45) $\xrightarrow{K_3Fe(CN)_6}$ (6.46) Purple (6.12)

exposure; this reagent converts the dyes in the layers into a form capable of diffusion. When, for example, the green-blue dye in the red-sensitized layer impinges on an exposed silver bromide grain the dye molecule is oxidized, which makes it resistant to diffusion so that it is retained in this layer; the purple and yellow dyes in the green- and blue-sensitive layers, respectively, behave similarly. The non-oxidized dye diffuses into a receiving layer where it is made resistant to diffusion, and forms a positive reproduction of the photographed object.

The 'Technicolor' process is also based on the principle of colour transfer (cf. references 6, 12). All classes of dyes can be used for these diffusion processes.

More recently silver-free colour photographic processes, in particular, are being investigated. Numerous articles—apart from the appropriate textbooks—have given a comprehensive explanation of all the problems of colour photography.[6,8,9]

6.6. LITERATURE

1. F. M. Hamer, 'The cyanines and related compounds,' in A. Weissberger, *The Chemistry of Heterocyclic Compounds*, Vol. 18, Interscience, New York, London, 1964.
2. H. Kuhn, *Helv. chim. Acta*, **31**, 1441 (1948); **32**, 2247 (1949); **34**, 2371 (1951); see also the summary: H. Kuhn, *Experientia (Basel)*, **9**, 41 (1953); N. S. Bayliss, *Quart. Rev. (chem. Soc. London)*, **6**, 319 (1952); J. R. Platt, *Free-Electron Theory of Conjugated Molecules*, University of Chicago Press, 1964. Also J. R. Platt, *J. chem. Physics*, **17**, 484, 1198 (1949).
3. H. Balli and F. Kersting, *Liebigs Ann. Chem.*, **647**, 1, 11 (1961); **663**, 96 (1963); H. Balli and R. Gipp, ibid., **699**, 133 (1966). Summary: S. Hünig, H. Balli, K. H. Fritsch, H. Herrmann, G. Köbrich, H. Werner, E. Grigat, F. Müller, H. Nother and K.-H. Oette, *Angew. Chem.*, **70**, 215 (1958); S. Hünig, H. Balli, E. Breither, F. Brühne, H. Geiger, E. Grigat, F. Müller and H. Quast, ibid., **74**, 818 (1962); S. Hünig and H. Balli, (a) *Chimia*, **22**, 147 (1968), (b) *Textilveredlung*, **4**, 37 (1969).
4. O. Isler, H. Lindlar, M. Montavon, R. Rüegg and P. Zeller, *Helv. chim. Acta*, **39**, 249 (1956); O. Isler, R. Rüegg and P. Schudel, *Chimia*, **15**, 208 (1961).
5. E. Klein and R. Matejec, *Photogr. Korresp.*, **101**, 3 (1965); H. Meier, *Spectral Sensitization.*, Focal Press, London, New York, 1968; see also ref. 8(a).
6. H. Ulrich and E. Klein, in *Ullmann's Encyclopädie der technischen Chemie*, 3rd edition, Urban and Schwarzenberg, München, 1962, Vol. 13, 603.
7. O. Reister, in *Ullmann's Encyclopädie der technischen Chemie*, 3rd edition, Urban and Schwarzenberg, München, 1963, Vol. 14, 310.
8. C. E. K. Mees and T. H. Jones; *The Theory of the Photographic Processes*, 3rd edition, The Macmillan Co., New York, 1966, (a) Chap. 11 and 12, (b) Chap. 17.
9. Summary: J. Nys, *Chimia*, **22**, 115 (1968).
10. M. Schellenberg and R. Steinmetz, *Helv. chim. Acta*, **52**, 431 (1969).
11. G. Crawley, *Brit. J. Photogr.*, **109**, 254 (1962); **110**, 76, 443 (1963).
12. P. N. James, *J. Soc. Motion Picture Television Engr.*, **74**, 989 (1965).

7

Aza[18]annulenes

THE chromophore of the dyes of the [18]annulene series† of ring systems is a cyclic system of conjugated double bonds, with 18 π-electrons. Their best-known representatives are the biologically important colouring matters of blood (haemin: $Fe^{3\oplus}$ complex of (7.1): $X = CH$, $R^1 = CH=CH_2$; $R^2 = CH_3$; $R^3 = CH_2—CH_2—COOH$) and leaves (chlorophyll A: (7.2)), and the synthetic phthalocyanine dyes; the former are derivatives of porphin ((7.1): $X = CH$; $R^1 = R^2 = R^3 = H$) and the latter derivatives of tetrazaporphin ((7.1): $X = N$; $R^1 = R^2 = R^3 = H$).

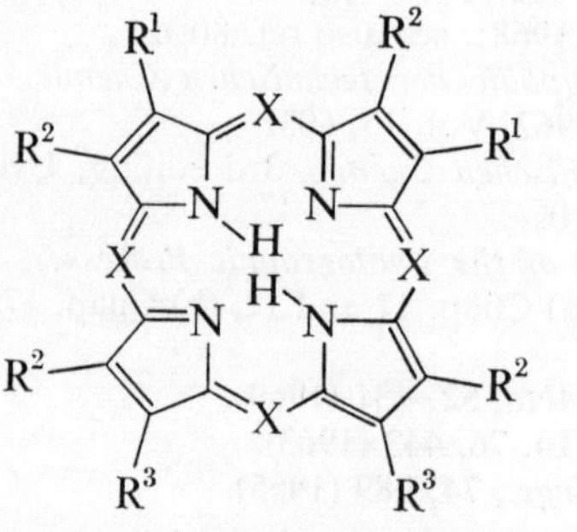

(7.1)

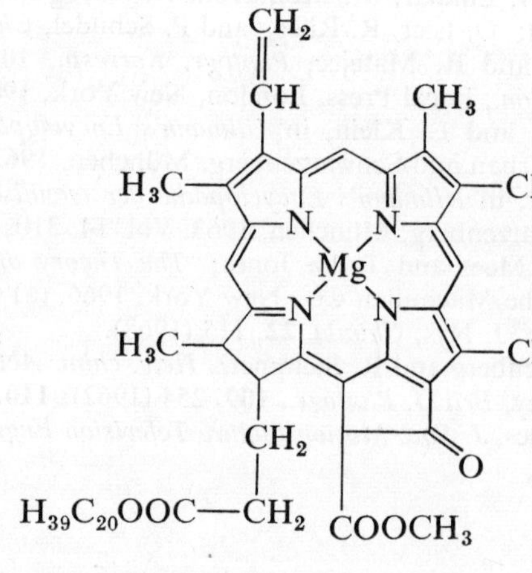

(7.2)

7.1. GENERAL REMARKS AND STRUCTURES

As F. Sondheimer has demonstrated,[1] not only cyclic hydrocarbons with n rings and $(4n + 2)$ conjugated π-electrons ($n = 1, 2, 3\ldots$), in accordance with the Hückel MO theory (cf. section 2.1), but also compounds with $(4n + 2)$ conjugated π-electrons in the perimeter ($n = 0, 1, 2\ldots$) assuming that the steric conditions permit a planar or approximately planar ring system have an aromatic character. Moreover the delocalization energy liberated in the aromatization of the many-membered ring has to be larger than that of the anellated rings. These criteria of

† The number in brackets corresponds to the number of π-electrons in the perimeter.

aromaticity not only are valid for carbocyclic compounds but also are of significance, with their aza analogues, as a comparison of dehydrophthalocyanine (7.3), phthalocyanine (7.4) and the aza[18]annulene (7.5) shows. (7.3) possesses in the central, many-membered ring a number of π-electrons, namely 16, which does not conform to the Hückel rule and it is therefore a nonaromatic, moderately stable, faintly yellow, compound. (7.4) has a conjugated system of 18 π-electrons and is a very stable blue compound. It must be emphasized however that it is very difficult to give a general definition of 'aromaticity'.[1]

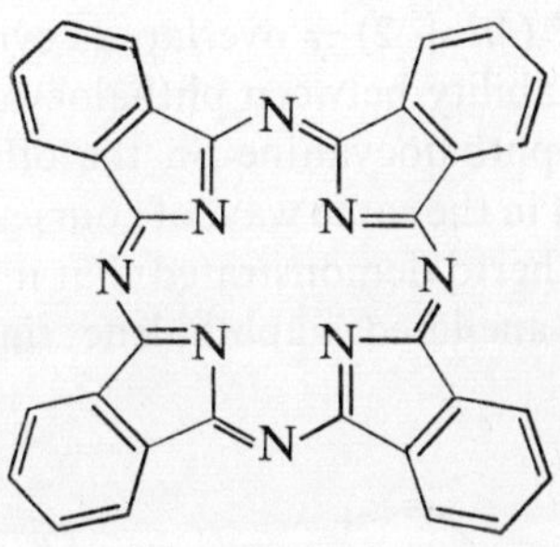

(7.3)

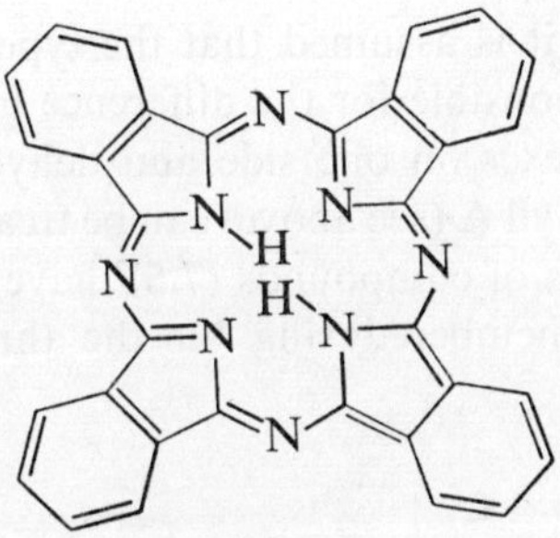

(7.4)

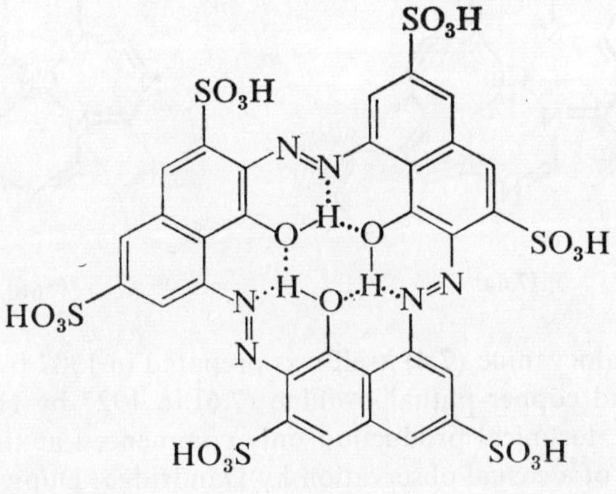

(7.5)

Like benzene the 18-membered ring is capable of maintaining a current induced by an outer magnetic field in the π-electron system and this is reflected, e.g. in the n.m.r. spectrum,[2] which provides evidence of aromaticity. Structures (7.6a) and (7.6b) demonstrate a specific problem in aza-annulene aromaticity. Hückel's rule (section 2.1) says that ring systems with $(4n + 2)$ π-electrons in the perimeter are aromatic; it does not say, however, how many atoms there should be in the perimeter. Formula (7.6a) shows, in bold bonds, a ring system of 18 atoms and 18 π-electrons. The heavy bonds in formula (7.6b) belong to a ring with 16 atoms only; in that ring, there are 14 π-electrons, one in each of the respective orbitals of the 8 carbon atoms and in 6 of the 8 nitrogen atoms; two nitrogen atoms 'contribute' however two lone pairs of p-electrons each in their orbitals of the overlapping π-orbital system; the result is a total of 18 π-electrons on 16 atoms.† In that mesomeric structure given in (**7.6b**) the nitrogens of the pyrrole nuclei in the upper left and the lower right corner contribute formally these 4 p-electrons. It is clear that a corresponding mesomeric structure can be written in which the other pyrrole has that function. Similarly for (**7.6a**) other 18-atom perimeters can be written.

Generally it is assumed that the type of $(4n + 2)$ π overlap as symbolized by (**7.6b**) is responsible for the difference in stability between phthalocyanine and its metal complexes on one side and dehydrophthalocyanine on the other. Haemin and chlorophyll A (see above) can be treated in the same way, of course. In contrast, investigations of compounds (**7.5**)[3] have hitherto demonstrated that it is obviously not the 18-membered ring but the three anellated naphthalene rings that are aromatic.

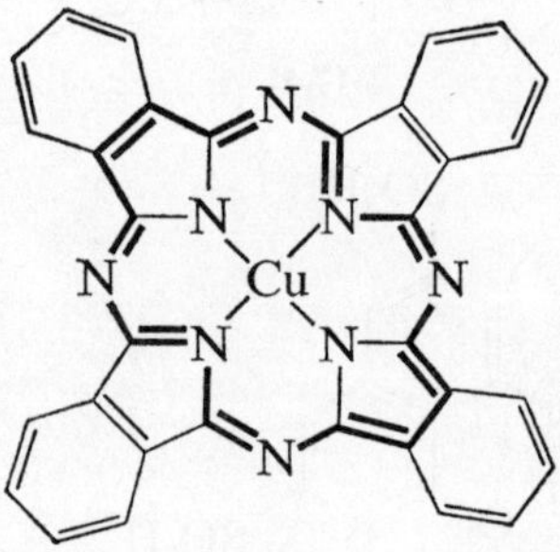

(**7.6a**) 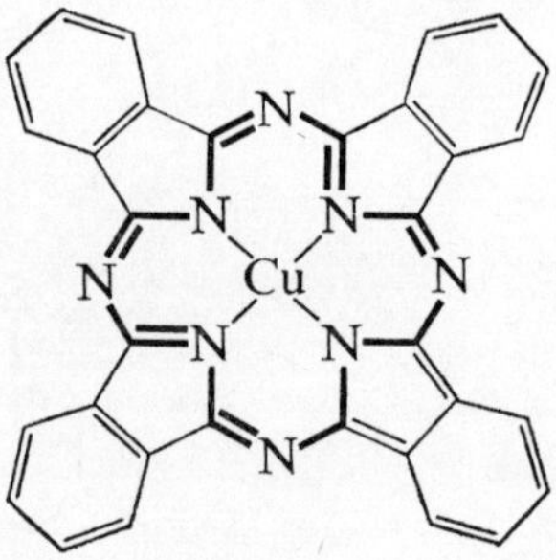   (**7.6b**)

Although phthalocyanine (**7.4**) itself was prepared in 1907 by A. von Braun and J. Tscherniak, and copper phthalocyanine (**7.6**) in 1927 by H. de Diesbach and E. von der Weid, technical production only commenced at the beginning of the 1930s, as a result of a casual observation by Dandridge, Dunworth, Drescher and Thomas of Scottish Dyes Ltd. (part of I.C.I.), in the course of the preparation of phthalimide from phthalic anhydride and ammonia.

The structure for phthalocyanine proposed by R. P. Linstead in 1933[4] was confirmed in 1935 by J. M. Robertson by means of X-rays.[5] The solubility characteristics of phthalocyanines depend markedly on the type of substitution and the coordinatively bound metal ion. The most important unsubstituted phthalocyanines, as for instance (**7.4**) and (**7.6**), as well as nickel and cobalt complexes of (**7.4**), dissolve readily in concentrated sulphuric acid but are practically insoluble in

† See footnote on p. 92.

organic solvents of low boiling point; they can be recrystallized in small amounts from some high-boiling solvents. In contrast to Na-, K-, Ca-, Ba-, Mg-, and Cd-complexes, Cu-, Zn-, Fe-, Co- and Pt-complexes of (**7.4**) will sublime *in vacuo* at 550–600°C without decomposition and are very resistant to acids (high complex-forming constants; cf. section 3.6). As X-ray and spectrophotometric (infrared) investigations have demonstrated, numerous phthalocyanine compounds exist as different polymorphous modifications; they play a significant part in the production of pigments (cf. section 7.3); so far, three crystal modifications of (**7.4**) (α-, β-, γ-forms) and two of (**7.6**) (a metastable α- and a stable β-form) are known. The long-wave absorption bands of the whole range of phthalocyanine dyes lie, without exception, in the blue spectral region (e.g. (**7.6**): λ_{max} (vapour) = 678 nm; log ε = 5·34). Changes of hue towards green are obtained by substitution in the benzene ring (cf. (**7.7**): W, X, Y, or Z = Cl, SH or phenyl), whilst replacement of the benzene rings gives rise to reddish derivatives (cf. (**7.8**) or the nickel complex of (**7.1**): $R^1 = R^2 = R^3$ = alkyl).

(7.7)   (7.8)   (7.9)

Apart from their application as dyes (cf. section 7.3) different metallic phthalocyanines, especially those of the $Fe^{3\oplus}$ complexes, can be used as oxidation catalysts[6] and they possess interesting semi-conductor properties.[7]

7.2. PRINCIPLES OF PREPARATION

The synthesis of the most important phthalocyanine derivatives usually succeeds with simple starting materials without isolation of intermediate products and they are generally obtained in good yield. The progress of the cyclization as well as the stoichiometry of the 'direct' (technical) syntheses (source of hydrogen atoms) is still not clear. The course of the reaction formulated in the equation scheme (7.1) for the preparation of (**7.4**) and (**7.6**) is supported by the intermediates which have been established so far and it comprises all the most important technical manufacturing methods. The separate reaction steps have been discussed in detail.[8] Substituted phthalocyanines are obtained either by direct substitution in an existing molecule of (**7.6**) or by synthesis with substituted starting materials. Compounds with halogen, sulphonate, sulphonyl chloride, sulphonamide and chloromethyl groups are preferably prepared by the first, i.e. the direct substitution method. The distribution of electron density, calculated according to the MO–LCAO method[9] (cf. section 2.1), suggests that electrophilic substituents first attack at positions 4 and 5 whilst nucleophilic reagents attack the non-benzenoid carbon atoms preferentially.

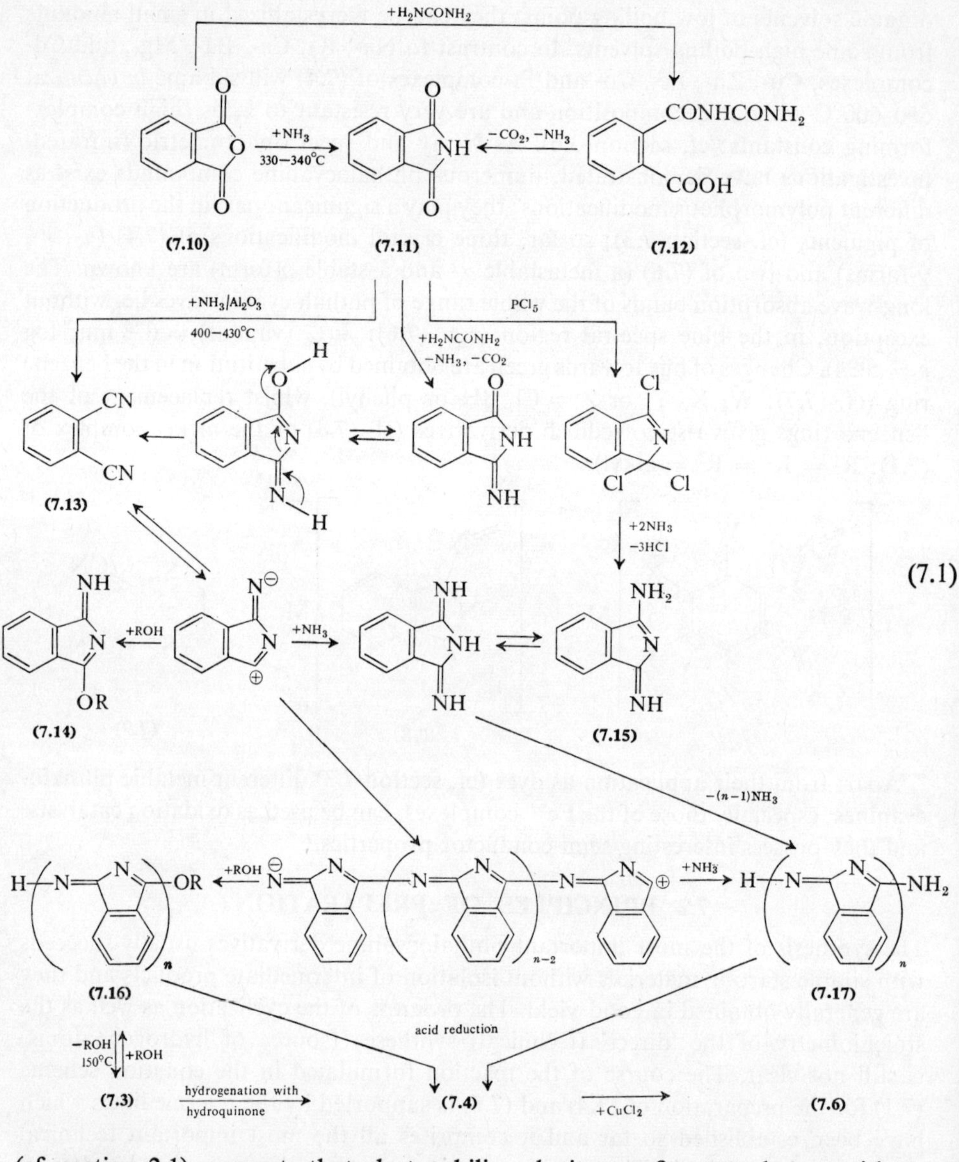

Owing to the oxidizing action of nitric acid, nitro groups cannot be introduced directly but only by synthesis from 4-nitrophthalimide. Because primary amino groups are not stable in the melt during synthesis, phthalocyanine derivatives containing such groups are prepared almost exclusively by reduction (e.g. with

Na_2S) of the corresponding nitrophthalocyanines. Similarly, mercapto groups are introduced indirectly by the reduction of sulphonyl chloride derivatives. With the exception of perchlorinated derivatives, all technical phthalocyanine substitution products as well as those compounds which are synthesized from monosubstituted starting materials are mixtures of position isomers; because of their symmetrical molecular structure the positions 3 and 6, and 4 and 5, respectively, are identical.

By analogy with the methods of making phthalocyanines the corresponding tetrazaphorphins can be obtained from the *o*-dinitriles and *o*-dicarboxylic acids of heterocyclic or condensed aromatic compounds. Among the cycloaliphatic *o*-dinitriles, 2,3-dicyan-1,4-dithiacyclohexane(2,3) (**7.9**) is nowadays of technical importance as starting component for the manufacture of tetrazaporphin with anellated dithia rings. It is easily prepared from sodium cyanide and carbon disulphide in dimethylformamide.

7.3. TECHNICAL APPLICATION OF AZA[18]ANNULENE DYES

Because of their good fastness properties phthalocyanines are suitable for almost all processes in which pigments are used (cf. section 11.2). The type and size of the crystals is of decisive importance in this application. Thus all production methods for (**7.6**) (Monastral Fast Blue B: C.I. Pigment Blue 15) yield coarse crystals of the stable β form which is unsuitable as a pigment. By various methods they can be converted into the metastable fine-particle α form (e.g. precipitation from H_2SO_4 or grinding with NaCl); when slightly moistened with toluene or xylene this is transformed into the β form, without enlargement of the crystals, and is then suitable for use as a pigment. Similarly (**7.4**) (Monastral Fast Blue G: C.I. Pigment Blue 16) has at least three crystal modifications. The green pigment Monastral Fast Green G (C.I. Pigment Green 7 (**7.7**): $W = X = Y = Z = Cl$; $Me = Cu^{2\oplus}$) is of great technical importance. Partially-sulphonated derivatives of (**7.6**) are commercially available as brilliant turquoise acid and direct dyes (two sulphonate groups per molecule: Chlorantine Fast Turquoise Blue GLL: C.I. Direct Blue 86). The reaction of the polysulphonylchloride of (**7.6**) with primary alkylamines which also contain tertiary amino groups in the alkyl residue yields strongly basic fast-to-light dyes which are soluble in dilute acetic acid as cationic compounds (e.g. Astra Blue Base 6GLL; C.I. Solvent Blue 51). When salicylic acid, substituted with amino or hydroxy groups, (instead of primary alkylamines) is used, chrome dyes are obtained. The detrimental effect of the water solubilizing groups on washing resistance can be eliminated either by the introduction of reactive groups which form a covalent bond with the substrate, or by special methods of application; thus the introduction of mercapto groups gives water-soluble polymercaptophthalocyanines (e.g. Thionol Ultra Green B: C.I. Sulphur Green 14; (**7.7**): $W = Y = Z = H$; $X = SH$; $Me = Cu^{2\oplus}$), which are adsorbed by the cellulose fibres and subsequently converted by atmospheric oxidation, or by $K_2Cr_2O_7$, to insoluble pigments (cf. Chapter 9). Phthalocyanines can also be used in the form of vat dyes. Indanthren Brilliant Blue 4G (C.I. Vat Blue 29; cobalt (III)-phthalocyanine monosulphonic acid) is an interesting commercial dye; it is difficult to dissolve as a zwitterion but can be converted to the adequately-soluble anionic cobalt (II)-complex by reduction (with $Na_2S_2O_4$) and dyed as such; when the fibres are subsequently exposed to air the relatively insoluble cobalt (III) compound is re-formed. Reactive groups (cf. section 4.2.6) are linked either indirectly (as e.g. the chlorinated heteroaromatics) with diamine (ethylenediamine, phenylenediamine) via a sulphonamide group, or directly (as e.g. vinylsulphone groups) to the phthalocyanine molecule. Derivatives of (**7.6**) with water solubilizing groups which will readily split off in the dyeing process have also attained technical importance: thus isothiuronium salt groups ($—\overset{\oplus}{S}=C[N(CH_3)_2]_2$) can be introduced by replacing the chlorine in polychloromethyl derivatives of (**7.6**) with tetramethylthiourea residues, which are split off during dyeing and printing by treatment with $K_2Cr_2O_7$ and acetic acid or by heating in the presence of dilute alkalis, e.g. Alcian Blue 8GX (C.I. Ingrain Blue 1). Furthermore, derivatives are commercially available which contain thiosulphonic acid groups, which also decompose during the dyeing process, yielding wash-fast turquoise shades on cotton. The relatively smooth formation of phthalocyanine and its metal complexes has led to an assortment of Phthalogen dyes, which are used principally as developing dyes for printing on cotton: the printing paste contains 1-amino- or 1-alkoxy-iminoisoindolenine ((**7.17**) or (**7.16**)) or their substitution products in a suitable solvent (glycols, dimethylformamide, triethanolamine) and a metal salt. The dye is formed in the fibre by steaming, or dry heat at 120–150°C. Phthalogen Blue Black IVM (C.I. Ingrain Blue 7) is a typical example; its printing paste contains a nickel salt and a mixture of 1-amino-3-iminoisoindolenines which are obtained from phthalodinitrile (**7.13**) and dithiacyclohexene dinitrile (**7.9**).

7.4. LITERATURE

1. F. Sondheimer, *Proc. Roy. Soc.* (*London*), **A297**, 173 (1967).
2. A. B. P. Lever, *Advances inorg. Chem. Radiochem.*, **7**, 27 (1965).
3. Z. J. Allan and J. Podstata, *Collect. czechoslov. chem. Commun.*, **34**, 282 (1969).
4. R. P. Linstead, *J. chem. Soc.* (*London*), **1934**, 1016; *Ber. dtsch. chem. Ges.*, **72A**, 93 (1939).
5. J. M. Robertson, *J. chem. Soc.* (*London*), **1935**, 615; **1936**, 1195, 1736; **1937**, 219; **1940**, 36.
6. H. Kropf, *Liebigs Ann. Chem.*, **637**, 73, 93, 111 (1960).
7. H. Meier, *Z. Elektrochem.*, **58**, 859, 867 (1954), also ref. 6, p. 76.
8. H. Vollmann, in *Ullmann's Encyclopädie der techn. Chemie*, 3rd edition, Urban and Schwarzenberg, München, 1962, Vol. 13, 696; cf. also F. H. Moser and A. L. Thomas, *Phthalocyanine Compounds*, Reinhold, New York, 1963.
9. S. C. Mathur, *J. Chem. Physics*, **45**, 3470 (1966).

8

Di- and Triarylcarbonium Dyes and their Aza Analogues

THIS Chapter describes a group of arylogue† poly(aza)methine dyes (see also Chapter 6), which can be regarded as resonance hybrids of several mesomeric structures (8.1).‡

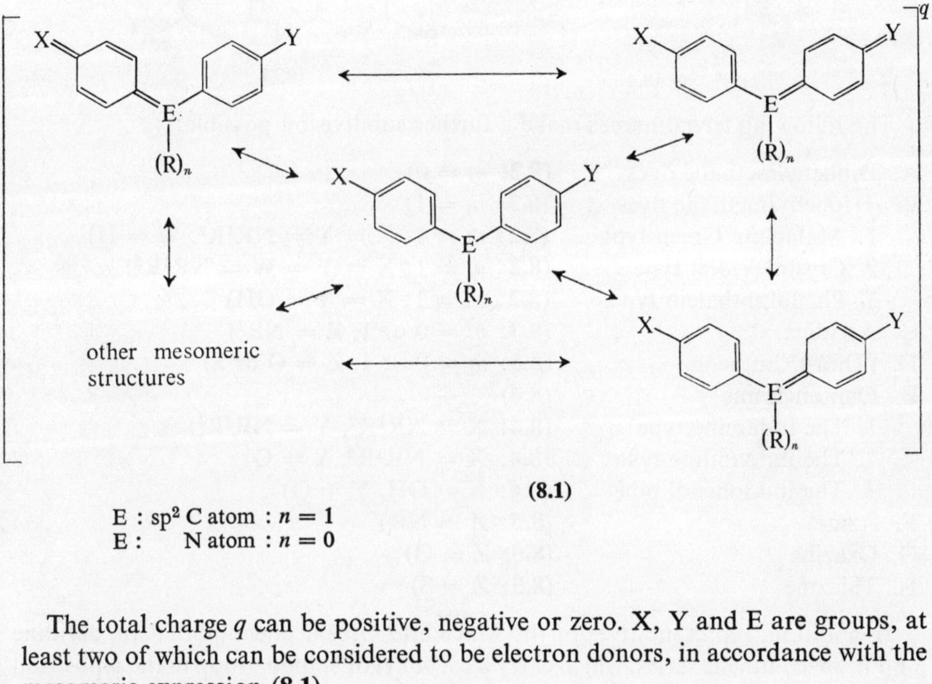

(8.1)

E : sp² C atom : $n = 1$
E :   N atom : $n = 0$

The total charge q can be positive, negative or zero. X, Y and E are groups, at least two of which can be considered to be electron donors, in accordance with the mesomeric expression (8.1).

8.1. GENERAL REMARKS AND STRUCTURES

All the important representatives of this class of dyes can be derived from the structures (8.2)–(8.5).

† We use the expressions arylogue and phenylogue respectively in analogy to the term vinylogue to indicate that electronic effects are transmitted through aromatic systems.
‡ For the significance of mesomeric structures, see section 2.1. In each of the following formulae only one of the mesomeric structures is shown.

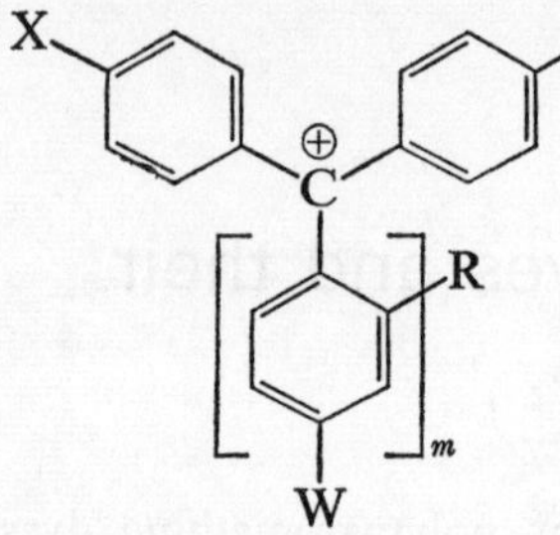

[Structures (8.2), (8.3), (8.4), (8.5)]

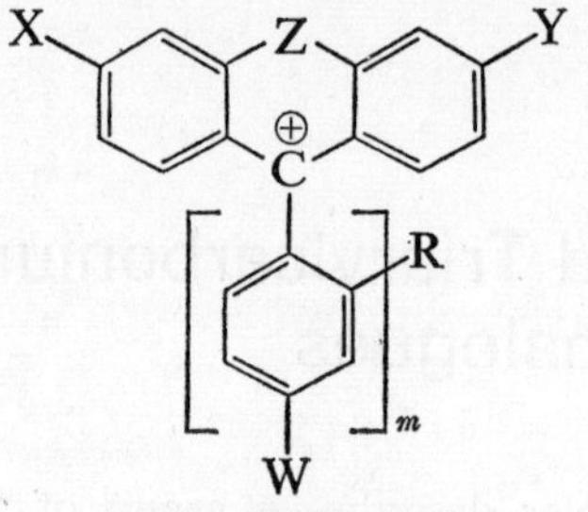

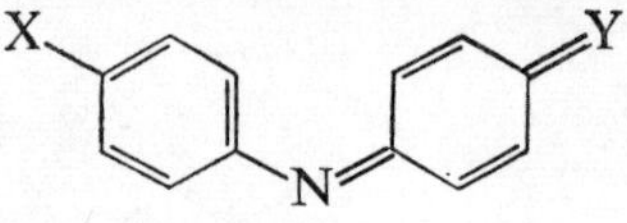

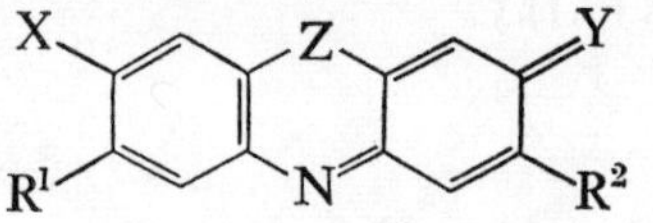

The following trivial names make a further subdivision possible:

A. Diphenylmethane dyes   (8.2: $m = 0$)
B. Triphenylmethane dyes   (8.2: $m = 1$)
   1. Malachite Green type   (8.2: $m = 1$; $X = Y = NR^1R^2$, $W = H$)
   2. Crystal Violet type   (8.2: $m = 1$; $X = Y = W = NR^1R^2$)
   3. Phenolphthalein type   (8.2: $m = 1$; $X = Y = OH$)
C. Acridine   (8.3: $m = 0$ or 1, $Z = NR^1$)
D. (Thio-)Xanthene   (8.3: $m = 0$ or 1, $Z = O$ or S)
E. Quinoneimine   (8.4)
   1. The indamine type   (8.4: $X = NR^1R^2$, $Y = \overset{\oplus}{NR^3R^4}$)
   2. The indoaniline type   (8.4: $X = NR^1R^2$, $Y = O$)
   3. The indophenol type   (8.4: $X = OH$, $Y = O$)
F. Azine   (8.5: $Z = NR$)
G. Oxazine   (8.5: $Z = O$)
H. Thiazine   (8.5: $Z = S$)

In addition, representatives of the Malachite Green, phenolphthalein, acridine and (thio-)xanthene series, which carry a carboxyl or sulpho group in the *o*-position to the central carbon atom ((8.2) and (8.3): $R = COOH$ or SO_3H respectively), are generally designated as phthaleins or sulphophthaleins respectively. By analogy with the classification of polymethine dyes the arylcarbonium dyes and their aza analogues, depending on the type of substituent W, X and Y in the *p*-position at the central carbon or nitrogen atom, could be regarded as phenylogue† (aza) stryptocyanines (cf. (6.2)), as phenylogue (aza)oxonoles (cf. (6.5) and (6.6)) or as phenylogue (aza)merocyanines (cf. (6.10)).

† See first footnote on p. 99.

Numerous characteristic reactions of these dyes are attributable to the electrophilic properties of the central carbon atom of (**8.2**) and (**8.3**), the ready nucleophilic substitution of W, X and Y (cf. (8.5)), the reversible formation of colourless carbinol bases and lactones (cf. (8.13)), as well as the generally moderate light fastness.[1] There are various reports on the X-ray structural analysis of di- and tri-arylcarbonium dyes.[2]

The first artificial dye, Mauveine ((**8.25**): $R^1 = CH_3$, $R^2 = $ phenyl), is nowadays only of historical interest. It was prepared in 1856 by W. H. Perkin by the oxidation of aniline containing toluidine with potassium bichromate in dilute sulphuric acid. Soon, further dyes followed which were all obtained by similar empirical methods, mainly by oxidation of a wide variety of aniline derivatives. The term 'aniline dyes' for all synthetic dyes obviously stems from that period.

Apart from the areas of application discussed in sections 8.3 and 8.4 the arylcarbonium dyes and their aza analogues are of some importance in colour photography (cf. section 6.5) and in medicine. Well-known examples are, among others, Mercurochrome, used for disinfecting wounds ((**8.27**): $R^1 = H$, $R^2 = R^4 = Br$, $R^3 = Hg—OH$) and Atebrin, an acridine derivative which is used for the prevention and treatment of malaria.

8.2. PRINCIPLES OF PREPARATION

The synthesis of arylcarbonium dyes proceeds stepwise from mono- via di- to triarylcarbonium compounds. The principle is briefly summarized in equations (8.1)–(8.4).

$$\begin{array}{c} R^1 \\ \diagdown \\ R^2 \diagup C \diagdown R^4 \end{array} \rightleftarrows \begin{array}{c} R^1 \quad R^3 \\ \diagdown \diagup \\ C_\oplus \\ R^2 \diagup \end{array} + R^{4\ominus} \qquad (8.1)$$

(**8.6**)   (**8.7**)

$$\begin{array}{c} R^1 \\ \diagdown \\ C=O(S) + H^\oplus \\ R^2 \diagup \end{array} \rightleftarrows \begin{array}{c} R^1 \\ \diagdown \\ C^\oplus\!\!—OH(SH) \\ R^2 \diagup \end{array} \qquad (8.2)$$

(**8.8**)   (**8.9**)

Compounds (**8.7**) and (**8.9**) containing a central carbon atom as electrophilic reaction centre are reacted with an aromatic nucleophilic compound (**8.10**) (8.3). This reaction can be repeated up to three times provided that the central carbon atom remains sufficiently electrophilic. The number of hydrogen atoms attached to the carbon of the intermediate product (**8.6**) or (**8.8**) indicates the number of oxidation steps necessary for the production of a triarylcarbonium dye (8.4). The most widely used electrophilic reagents are, for example, phosgene ((**8.8**): $R^1 = R^2 = Cl$),

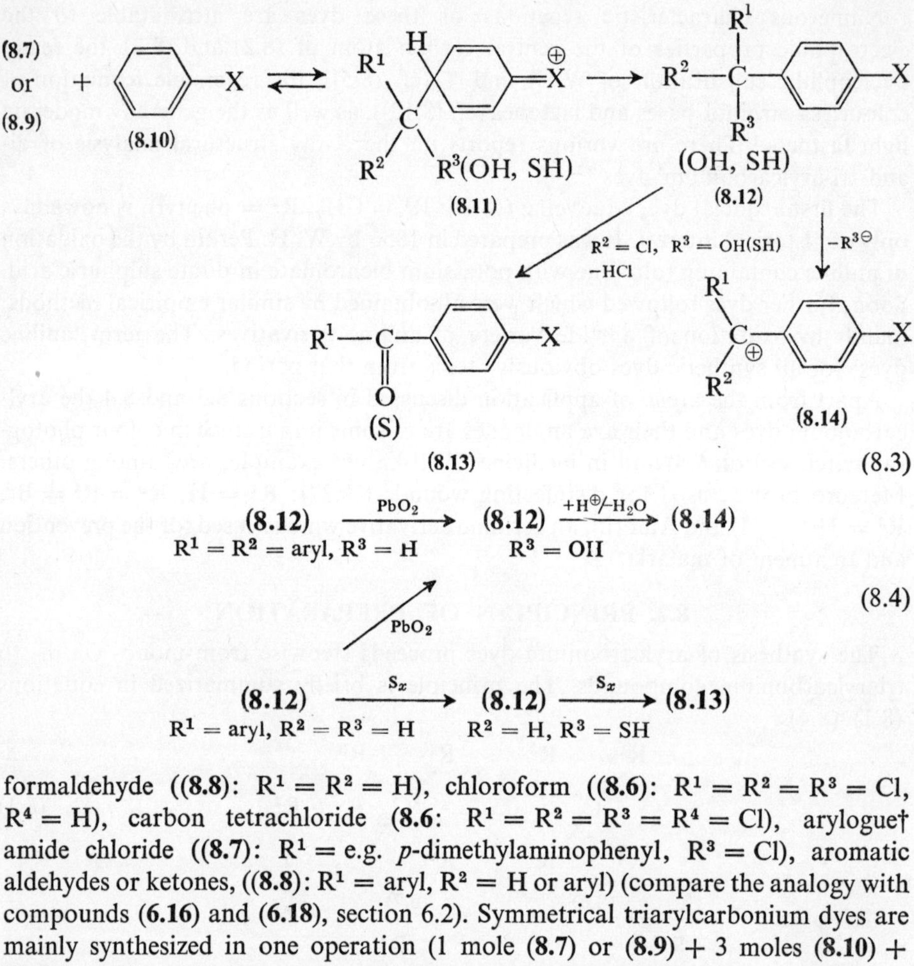

(8.7) or (8.9) + ![benzene-X] (8.10) ⇌ R^1–CH–[cyclohexadiene]=X$^\oplus$ / R^2 R^3(OH, SH) (8.11) → R^2–C(R^1)(R^3)–[benzene]–X (OH, SH) (8.12)

R^2 = Cl, R^3 = OH(SH) −HCl ↙   ↓ $-R^{3\ominus}$

R^1–C(=O)–[benzene]–X (S) (8.13) R^1–C$^\oplus$(R^2)–[benzene]–X (8.14)

(8.3)

$$(8.12) \xrightarrow{PbO_2} (8.12) \xrightarrow{+H^\oplus/-H_2O} (8.14)$$
$R^1 = R^2$ = aryl, R^3 = H R^3 = OH

↗ PbO_2

(8.4)

$$(8.12) \xrightarrow{S_x} (8.12) \xrightarrow{S_x} (8.13)$$
R^1 = aryl, $R^2 = R^3$ = H R^2 = H, R^3 = SH

formaldehyde ((**8.8**): $R^1 = R^2$ = H), chloroform ((**8.6**): $R^1 = R^2 = R^3$ = Cl, R^4 = H), carbon tetrachloride (**8.6**: $R^1 = R^2 = R^3 = R^4$ = Cl), arylogue† amide chloride ((**8.7**): R^1 = e.g. *p*-dimethylaminophenyl, R^3 = Cl), aromatic aldehydes or ketones, ((**8.8**): R^1 = aryl, R^2 = H or aryl) (compare the analogy with compounds (**6.16**) and (**6.18**), section 6.2). Symmetrical triarylcarbonium dyes are mainly synthesized in one operation (1 mole (**8.7**) or (**8.9**) + 3 moles (**8.10**) + PbO_2), whilst with the unsymmetrical dyes the intermediate products (**8.12**) are frequently isolated so that they can again react as (**8.7**) or (**8.9**) with another aromatic group (**8.10**).

It is therefore possible to prepare Crystal Violet ((**8.2**): W = X = Y = N(CH$_3$)$_2$, R = H, m = 1) in various ways. The direct path, namely reaction of phosgene or carbon tetrachloride with dimethylaniline gives firstly Michler's ketone (4,4'-tetramethyldiaminobenzophenone) via *p*-dimethylaminobenzoylchloride or -benzotrichloride or the phenylogue amide chloride (**8.14**: X = N(CH$_3$)$_2$, R^1 = *p*-dimethylaminophenyl, R^2 = Cl), and subsequently Crystal Violet. Alternatively, by reacting formaldehyde with dimethylaniline, 4,4'-tetramethyldiaminodiphenylmethane is obtained, which, on oxidation, yields Michler's

† See first footnote on p. 99.

hydrol ((**8.12**): $X = N(CH_3)_2$, $R^1 = p$-dimethylaminophenyl, $R^2 = H$, $R^3 = OH$) (cf. (8.4)). Further reaction with dimethylaniline converts this into the leucobase (**8.12**) ($X = N(CH_3)_2$, $R^1 = R^2 = p$-dimethylaminophenyl, $R^3 = H$), which may also be prepared by the condensation of 1 mole 4-dimethylaminobenzaldehyde with 2 moles dimethylaniline. Oxidation (cf. (8.4)) transforms the leuco compound into a colourless carbinol (**8.12**) ($X = N(CH_3)_2$, $R^1 = R^2 = p$-dimethylaminophenyl, $R^3 = OH$) which dissociates on acidification and produces the carbonium dye. In the presence of strong hydride ion acceptors such as $AlCl_3$ or chloranil (cf. section 3.5), the dye can also form directly from the leucobase.[3] Similarly, the unsymmetrical triphenylmethane dye, Malachite Green ((**8.14**): $X = H$, $R^1 = R^2 = p$-dimethylaminophenyl) can be prepared from 1 mole benzaldehyde, 2 moles dimethylaniline in the presence of an oxidant (usually PbO_2). In the synthesis of phthaleins (e.g. phenolphthalein (**8.30**)–(**8.35**)) the phthalic anhydride derivative is the electrophilic component; apart from this, their preparation conforms with the reaction scheme (8.1)–(8.4). An interesting variation of the synthesis is the so-called 'fusion' method: nucleofugal leaving groups (X^1), such as halogens, amino, alkoxy or sulphonic acid groups in the p-position to the central carbonium carbon, can be exchanged in nucleophilic substitution with aromatic amines (X^2) (8.5). The starting materials for xanthenes ((**8.3**): $Z = O$) and acridines ((**8.3**): $Z = NH$) are m-substituted phenols and m-substituted anilines (cf. (8.6)) respectively. Quinoneimine dyes (**8.4**) can be prepared in various ways; frequently a phenol or aniline derivative ((**8.10**): $X = NR^1R^2$ or OH) is condensed with a p-nitrosodialkylaniline ((**8.15**): $Y = NR^1R^2$) or a p-nitrosophenol ((**8.15**): $Y = OH$) (8.7).

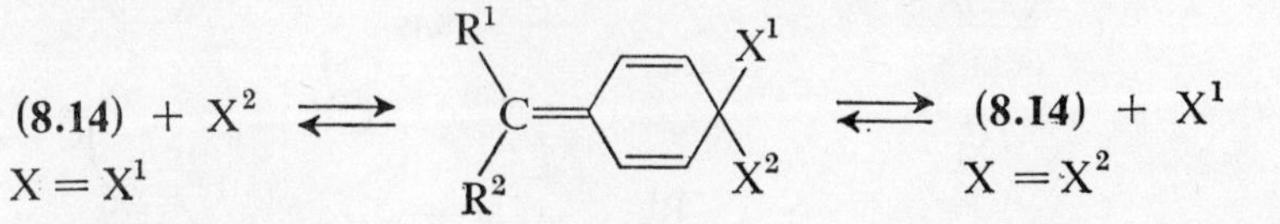

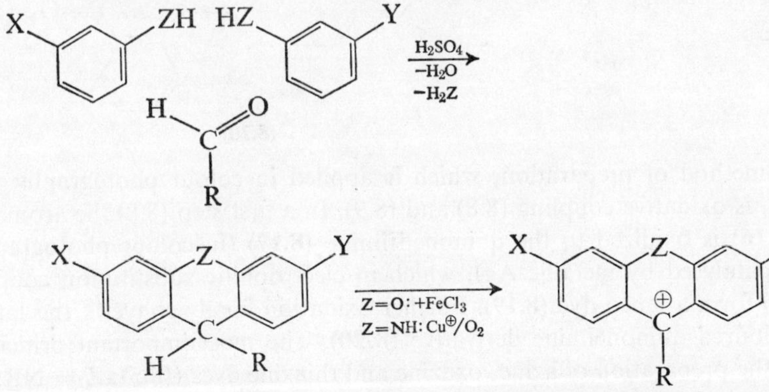

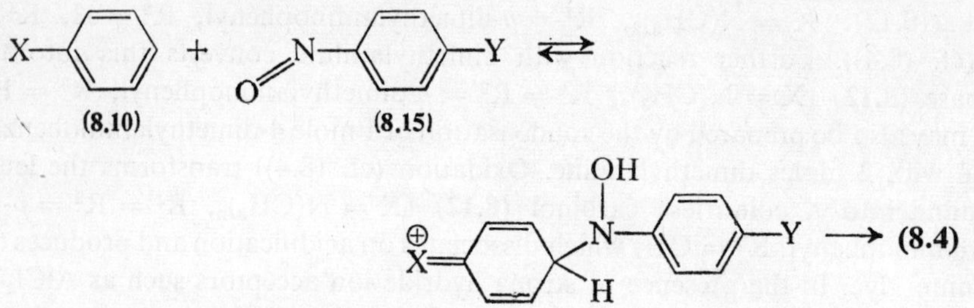

[Structures 8.10, 8.15, 8.4, 8.7, 8.16, 8.17, 8.8, 8.18, 8.19, 8.20]

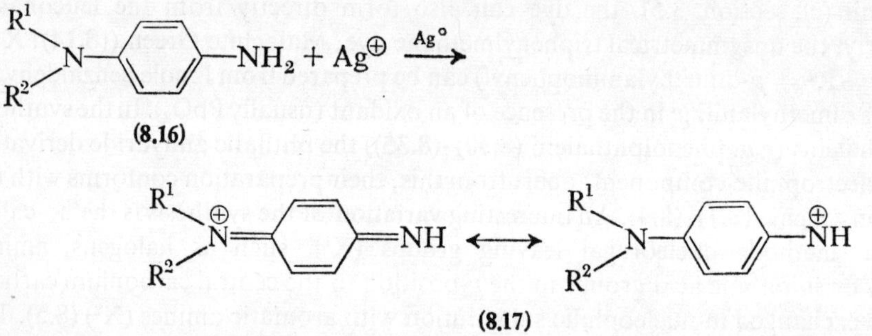

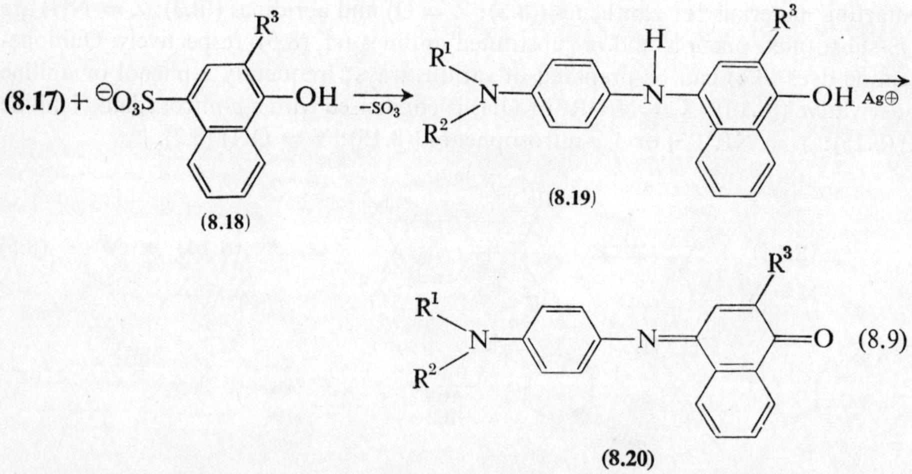

Another method of preparation, which is applied in colour photography (cf. section 6.5), is oxidative coupling (8.8) and (8.9). In a first step (8.8) the aromatic diamine (**8.16**) is oxidized to the quinonediimine (**8.17**) (in colour photography with $Ag^{\oplus}$, catalysed by metallic Ag), which in electrophilic substitution couples promptly to form a leuco-dye (**8.19**). Further oxidation finally converts the latter into the coloured quinoneimine derivative (**8.20**). The most important principle underlying the preparation of azine, oxazine and thiazine dyes ((**8.5**): Z = NR, O and S, respectively) are summarized in equations (8.10)–(8.12). Their production results mainly from oxidative ring closure of *o*-substituted quinoneimines (8.11).

The *o*-hydroxy and the *o*-amino compounds can be obtained by condensation (cf. (8.7)) of either the *o*-hydroxy (or *o*-amino) nitroso derivatives of (**8.15**) with (**8.10**), or the corresponding *m*-substituted derivatives of (**8.10**) with (**8.15**). Frequently amino groups are introduced only at the quinoneimine stage (8.10). The synthesis of thiazine often proceeds via the intermediate product (**8.22**) (ZR = S—SO$_3$H), which proceeds via the quinonediimino-thiosulphonic acid (**8.24**) (8.12).

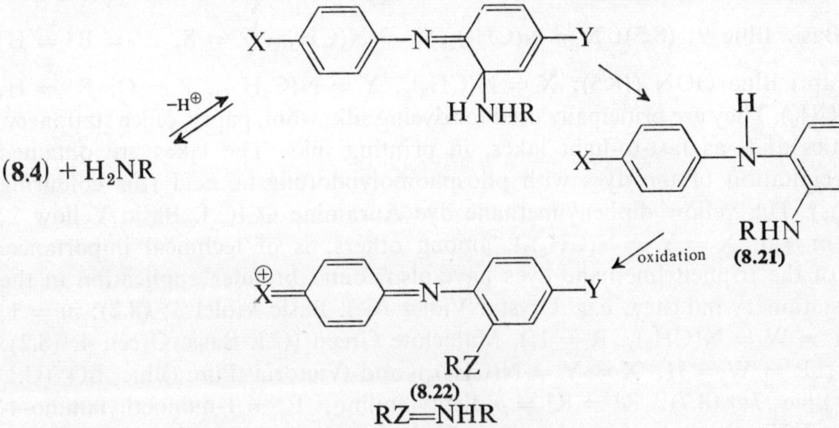

(**8.4**) + H$_2$NR

(8.10)

(**8.22**) ⇌

(8.11)

→ oxidation → (**8.5**)

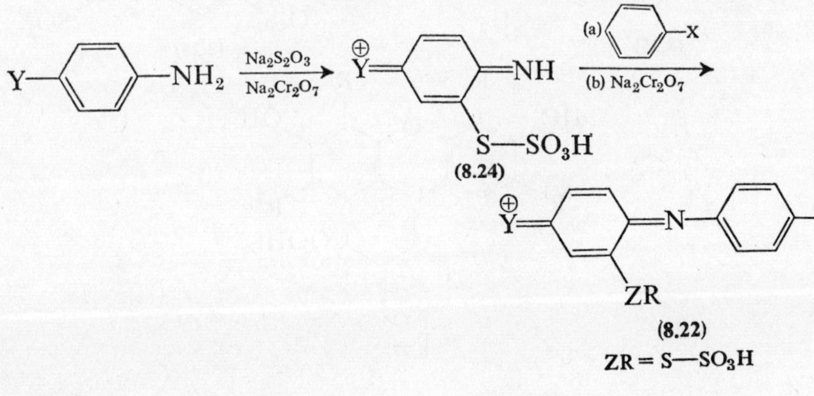

(8.12)

(**8.22**)
ZR = S—SO$_3$H

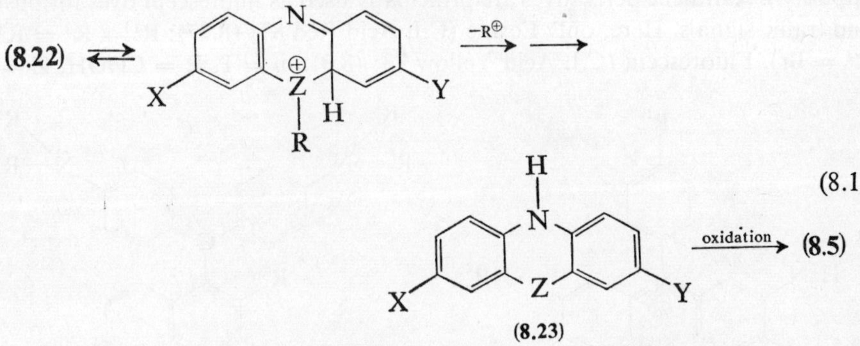

8.3. APPLICATION OF DI- AND TRIARYLCARBONIUM DYES IN DYEING

Dyes with the structures (8.2)–(8.5) are used only where there are no great demands on light fastness; the exceptions are cationic ('basic') compounds, which recently—owing to their relatively high light fastness on polyacrylonitrile—have again acquired some importance. A few well-known examples are Methylene Blue (C.I. Basic Blue 9; (8.5): $X = N(CH_3)_2$, $Y = \overset{\oplus}{N}(CH_3)_2$, $Z = S$, $R^1 = R^2 = H$) and Capri Blue GON ((8.5): $X = N(CH_3)_2$, $Y = \overset{\oplus}{N}(C_2H_5)_2$, $Z = O$, $R^1 = H$, $R^2 = CH_3$). They are principally used for dyeing silk, wool, paper, office stationery, cosmetics and, as fast-to-light lakes, in printing inks. The lakes are obtained by precipitation of the dyes with phosphomolybdotungstic acid (for colouring beacons). The yellow diphenylmethane dye Auramine O (C.I. Basic Yellow 2; (8.2): $m = 0$, $X = Y = N(CH_3)_2$), among others, is of technical importance. Some of the triphenylmethane dyes have also found broader application in the office stationery industry, e.g. Crystal Violet (C.I. Basic Violet 3; (8.2): $m = 1$, $X = Y = W = N(CH_3)_2$, $R = H$), Malachite Green (C.I. Basic Green 4; (8.2): $m = 1$, $R = W = H$, $X = Y = N(CH_3)_2$) and Victoria Pure Blue BO (C.I. Basic Blue 7; (8.7): $R^1 = R^2 = $ p-diethylanilino, $R^3 = $ 1-monoethylamino-4-naphthyl). Xanthene derivatives are principally used as fluorescent dyes for posters and traffic signals. Here, only Eosine (C.I. Acid Red 87; (8.27): $R^1 = R^2 = R^3 = R^4 = Br$), Fluorescein (C.I. Acid Yellow 73; (8.3): $m = 1$, $R = COOH$, $Z = O$,

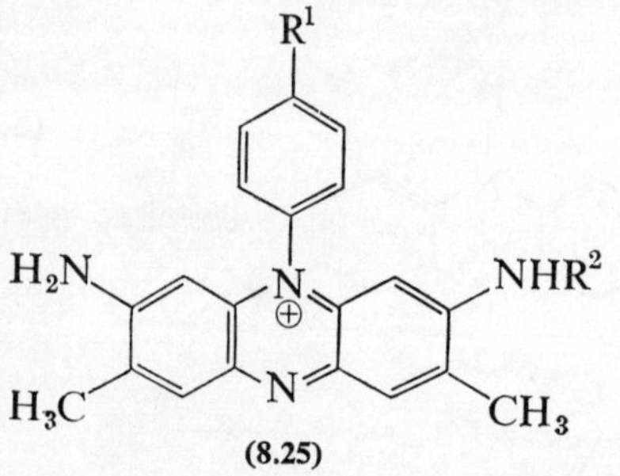

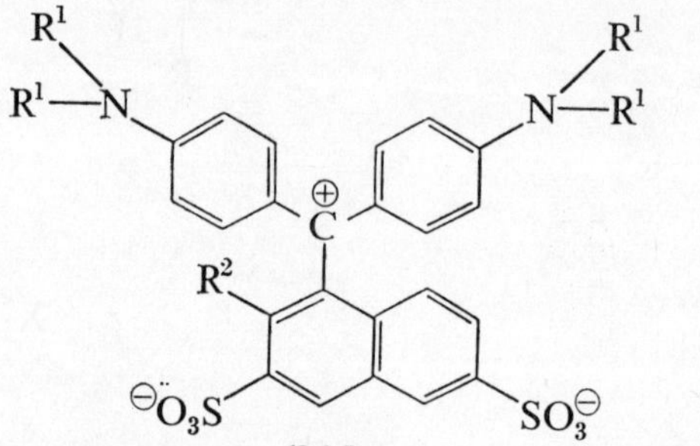

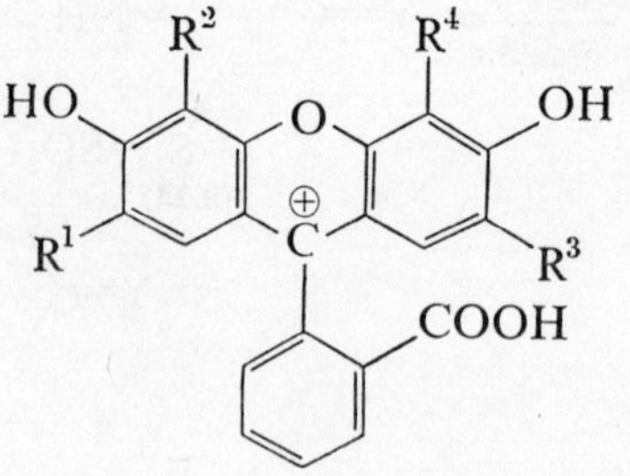

W = H, X = Y = OH) and Rhodamine B (C.I. Basic Violet 10; (**8.3**): $m = 1$, R = COOH, Z = O, W = H, X = Y = $N(C_2H_5)_2$) need to be mentioned. Whilst there is only historical interest in the simple quinoneimine derivatives such as Phenylene Blue ((**8.4**): X = NH_2, Y = NH_2) or Bindschedler's Green ((**8.4**): X = $N(CH_3)_2$, Y = $\overset{\oplus}{N}(CH_3)_2$), Fast Blue Z (C.I. Solvent Blue 22; (**8.20**): $R^1 = R^2 = C_2H_5$, $R^3 = H$) is still commercially available as an oil-soluble dye. Safranine T Extra (C.I. Basic Red 2; (**8.25**): $R^1 = R^2 = H$) is an important red paper dye which is related to Mauveine ((**8.25**): $R^1 = CH_3$, $R^2 = $ phenyl).

(**8.28**)   (**8.29**)

The introduction of at least two sulphonic groups produces anionic (acid) carbonium dyes which are used for dyeing wool, silk and leather. They include, for example, the two important green wool dyes, Wool Green BS (C.I. Acid Green 50; (**8.26**): $R^1 = CH_3$, $R^2 = OH$) and Naphthalene Green V (C.I. Acid Green 16; (**8.26**): $R^1 = C_2H_5$, $R^2 = H$). Some hydroxy triarylcarbonium compounds are commercially available; they are chrome dyes which are, among others, used for dyeing wool or polypropylene. They also contain a carboxyl group in the *o*-position to the hydroxyl group (complex-forming salicylic acid group, cf. section 3.6). Chrome Violet (C.I. Mordant Violet 39; (**8.7**): $R^1 = R^2 = R^3 = $ 3-carboxy-4-hydroxyphenyl) is a technical product. Aniline Black (**8.28**) is an interesting developing dye used for dyeing cotton. It is obtained by impregnating fabrics with aniline hydrochloride, sodium chlorate, ammonium vanadate and potassium ferrocyanide and then ageing at 60°C or steaming at 100°C.

Dioxazine compounds of the general structure (**8.29**) are nowadays used as violet pigments.

8.4. THE APPLICATION OF DI- AND TRIARYLCARBONIUM DYES AS INDICATORS

Colour indicators are substances which denote the point of equivalence of a chemical reaction by a change of colour. A small amount is added to the reaction mixture. The main spheres of application of indicators are, among others, acidimetry, measurement of redox potential and precipitation titrations. The principle underlying the action of colour indicators is explained in (8.13) by using the example of the acid–base indicator phenolphthalein: the acid and the lactone forms ((**8.30**), (**8.31**), (**8.33**), (**8.35**)) of this dye are colourless, but the alkaline form (**8.32**) is red;

in extremely alkaline conditions the colourless carbinol (**8.34**) finally forms. Neutralization of an acid or a base results in a pH jump at the point of equivalence; this change in pH causes the added indicator to change colour, signifying the endpoint of titration.

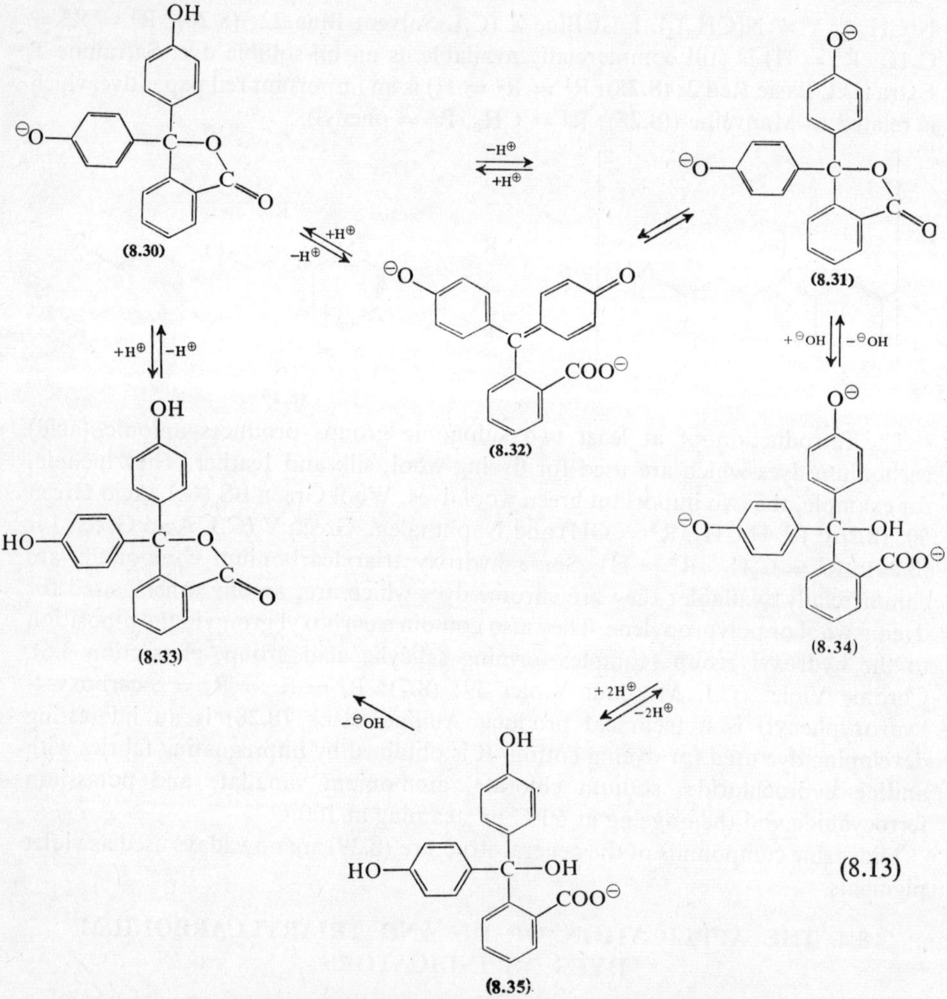

$$(8.13)$$

Central C atom: sp^2: red (**8.32**)
Central C atom: sp^3: colourless (**8.30, 8.31, 8.33–8.35**)

Analogous conditions exist with redox indicators; their use has been dealt with in a comprehensive article.[4]

Adsorption indicators are generally fluorescent dyes (e.g. xanthene dyes) which find an interesting application in precipitation titrations. They have a different colour in solution from that in the adsorbed state. For instance, when chloride ions are titrated with silver nitrate there is at first an excess of chloride which is adsorbed by the precipitated silver chloride giving it a negative charge. When the equivalence point is reached, a small excess of silver ions causes the silver chloride precipitate to be positively charged so that it adsorbs the anionic indicator dye from solution; the resultant change in colour signifies the endpoint of the titration.

8.5. LITERATURE

1. K. Iwamoto, *Bull. chem. Soc. Japan*, **10**, 420 (1935).
2. C. Stora and coworkers, *C.R. hebd. Séances Acad. Sci.*, **245**, 1626, 1693 (1957); **260**, 1660 (1965); *Bull. Soc. chim. France, Mém.*, **1966**, 841.
3. Summary: B. D. Tilak, *Chimia (Aarau, Schweiz)*, **20**, 272 (1966); C. D. Ritchie, W. F. Sager and E. S. Lewis, *J. org. Chemistry*, **84**, 2349 (1962).
4. J. N. Brazier and W. I. Stephen, *Analytica chim. Acta (Amsterdam)*, **33**, 625 (1965).

9
Sulphur Dyes

THE sulphur dyes are a class of water-insoluble, macromolecular, coloured compounds obtained by treating aromatic amines, phenols and aminophenols with sulphur and/or sodium polysulphide. By the action of sodium sulphide, these dyes are converted during the dyeing process into water-soluble derivatives which are adsorbed by cellulose fibres and are subsequently insolubilized by atmospheric oxidation.

9.1. GENERAL REMARKS

Sulphur dyes were introduced 70 years ago (R. Vidal 1893; L. Haas and R. Herz 1908). They are principally mixtures of coloured compounds and, because of their macromolecular structure and their insolubility in nearly all solvents, their constitutions have never been completely determined. Until now, in only one case has the degradation to low molecular weight derivatives, which could be identified by independent syntheses, been successful.[1]

9.2. STRUCTURES, AND PRINCIPLES OF PREPARATION

The starting products are sulphurized, either by heating, dry, with sulphur or sodium polysulphide (dry process dyes), or by boiling in a solvent (water, ethanol, n-butanol etc.) under reflux, or under pressure at temperatures up to about 150°C (solvent process dyes). The reaction conditions and type of reagent control the constitution, colour and fastness properties of the products. The most frequent structural component of dyes prepared by the dry process is the benzothiazole group whilst those prepared in solvents are largely derivatives of phenothiazones (**9.7**), phenazones and phenoxazones.

The principles of preparation are described in equations (9.1)–(9.5).

Sulphur and the sulphur-containing reagents (sodium polysulphide, sulphur monochloride) can act as reducing agents (nitro to amino group), oxidants, dehydrogenating agents (cf. (9.3)) and nucleophilic reagents (cf. (9.4) and (9.5)).

In order to produce technically useful sulphur dyes the substituents X, Y and Z in the quinonoid nucleus (cf. (9.3)–(9.5)) have to be either groups which can readily be subjected to nucleophilic substitution (e.g. halogen) or hydrogen. In contrast, any

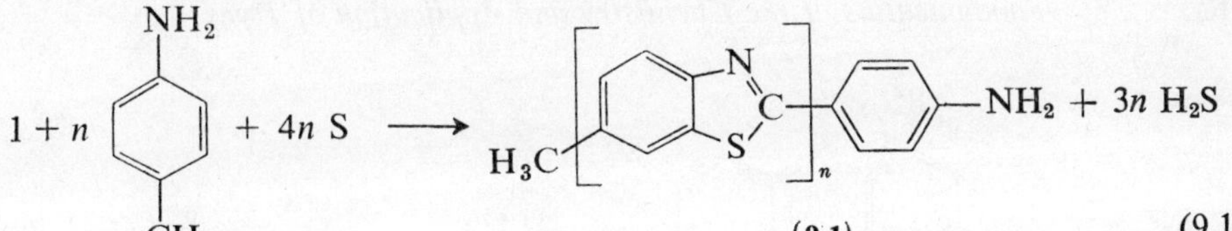

$$1 + n \; \underset{CH_3}{\underset{|}{C_6H_3}}\text{-}NH_2 + 4n\,S \longrightarrow \left[\text{benzothiazole}_n \right]\text{-}NH_2 + 3n\,H_2S \quad (9.1)$$

$n = 1$: Dehydrothiotoluidine
$n = 2$: Primulin base

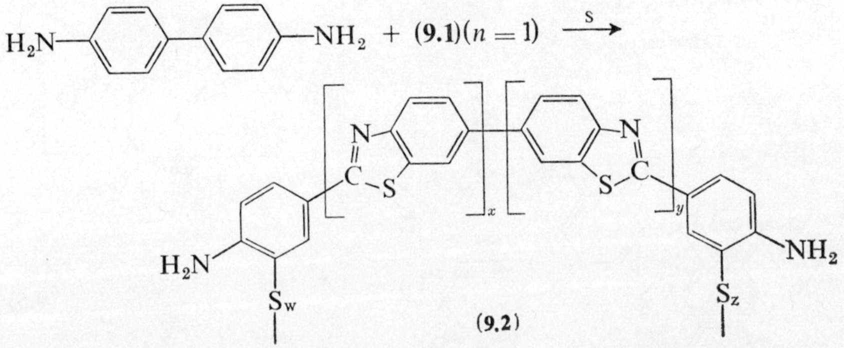

$$H_2N\text{-}C_6H_4\text{-}C_6H_4\text{-}NH_2 + (9.1)(n=1) \xrightarrow{S} (9.2) \quad (9.2)$$

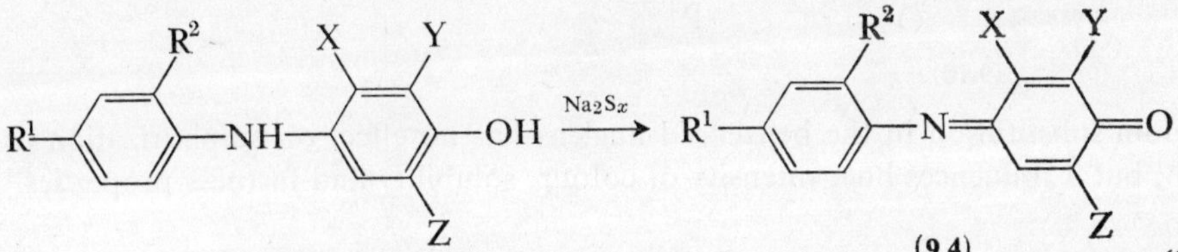

$$(9.3) \xrightarrow{Na_2S_x} (9.4) \quad (9.3)$$

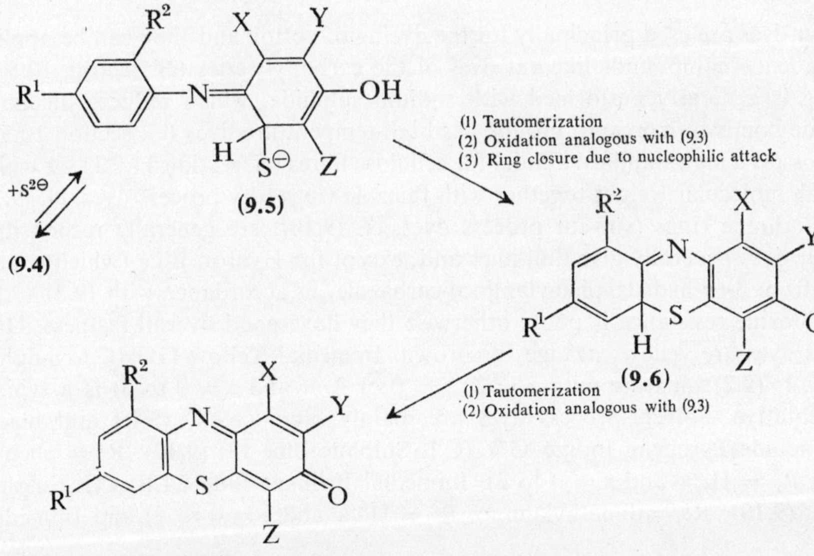

$$(9.4) \underset{+S^{2\ominus}}{\rightleftarrows} (9.5) \xrightarrow[\substack{(1)\;\text{Tautomerization}\\(2)\;\text{Oxidation analogous with (9.3)}\\(3)\;\text{Ring closure due to nucleophilic attack}}]{} (9.6) \xrightarrow[\substack{(1)\;\text{Tautomerization}\\(2)\;\text{Oxidation analogous with (9.3)}}]{} (9.7) \quad (9.4)$$

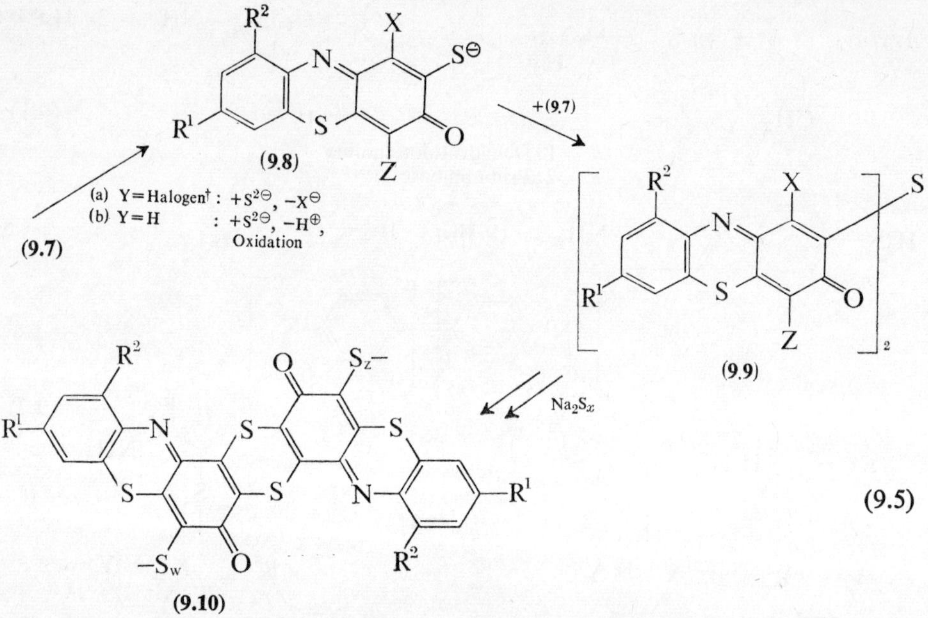

$$
\begin{array}{c}
\text{(a) } Y = \text{Halogen}^\dagger : +S^{2\ominus}, -X^\ominus \\
\text{(b) } Y = H \quad : +S^{2\ominus}, -H^\oplus, \\
\text{Oxidation}
\end{array}
$$

(9.7) → (9.8) →+(9.7)→ (9.9)

↙ Na₂Sₓ

(9.10)

(9.5)

random substitution in the benzenoid nucleus has no effect on sulphurization as such, but it influences hue, intensity of colour, solubility and fastness properties.[1]

9.3. APPLICATION OF SULPHUR DYES

Sulphur dyes are used principally for the dyeing of cotton and they can be applied via the leuco compounds like vat dyes of the carbonyl series (cf. section 10.5.5). Vatting is generally performed with sodium sulphide, which reduces the polysulphide bonds to mercapto groups. As high-temperature dyes (cf. section 10.5.5) they possess a high standard affinity for cellulose fibres (cf. section 11.3.1) for which the high molecular weight together with thiazole rings (dry process dyes, cf. (**9.2**)) or thianthrene rings (solvent process dyes, cf. (**9.10**)) are generally responsible. Sulphur dyes generally give dull hues and, except for Hydron Blue (which is produced from 3-(p-hydroxyphenylamino)-carbazole, in accordance with (9.3)–(9.5)) their chlorine resistance is poor; otherwise they have good overall fastness. Dry-process dyes are yellow, orange, or brown. Immedial Yellow GG (C.I. Sulphur Yellow 4; (**9.2**): mixture with $x = 2$, $y = 0$ or 2, w and $z = 0$ to 3) is a typical representative. Solvent-process dyes are mainly blue, green, violet and black. They include Pyrogene Indigo GW (C.I. Sulphur Blue 13; (**9.10**): R^1 = phenylamino, R^2 = H, w and $z = 0$ to 3), Immedial Brilliant Blue CLB (C.I. Sulphur Blue 9; (**9.10**): R^1 = dimethylamino, R^2 = H, w and $z = 0$ to 3) and Immedial

† Or other groups which can be readily subjected to nucleophilic substitution.

Black V (C.I. Sulphur Black 12; (**9.10**)): $R^1 = R^2 = NH_2$, w and $z = 0$ to 3). As regards quantity, Sulphur Black T (C.I. Sulphur Black 1) is by far the most important dye amounting to 10 per cent by weight of the total world production of dyes. It can be prepared by boiling 2,4-dinitrophenol with sodium polysulphide and is presumably also a phenothiazonethianthrene derivative (cf. (**9.10**)).

9.4. LITERATURE

1. W. Zerweck, H. Ritter and M. Schubert, *Angew. Chem.*, **A60,** 141 (1948).

10

Carbonyl Dyes

ALL dyes with at least two conjugated carbonyl groups are classified as carbonyl dyes **(10.1)**. These include indigo (section 10.1), anthraquinone (section 10.2), higher anellated and other carbonyl compounds (sections 10.3 and 10.4).

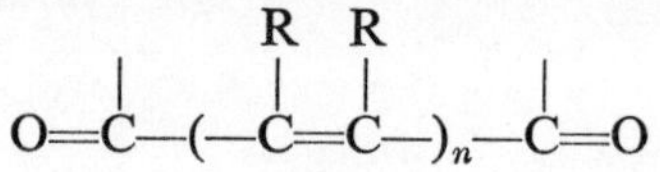

$$O=C-(-C=C-)_n-C=O$$
$$\text{with R substituents}$$
(10.1)

The great importance of the dyes in this class is mainly attributable to two facts: (a) their capacity, as conjugated dicarbonyl compounds, to be reduced to water-soluble dienols and being applied as such, gives them a special position in dyeing (vat dyes, section 10.5.5), and (b) long-wavelength absorption bands can be obtained with short conjugated systems, particularly with indigo and other anthraquinone derivatives. Thus, by relatively simple substitution with electron donors in the 1-, 1,4-, 1,5- and related positions of anthraquinone, compounds can be produced that absorb in any desired region of the visible spectrum. Whilst, for example, green azo dyes have a very complicated structure, 1,4-diphenylamino-anthraquinone is already green. This phenomenon is based on the so-called rule of distribution of auxochromes (cf. section 2.1).

10.1. INDIGO AND ITS DERIVATIVES

10.1.1. General remarks and structures

All indigo dyes contain the group **(10.2)** as the parent system, which is responsible for the colour.

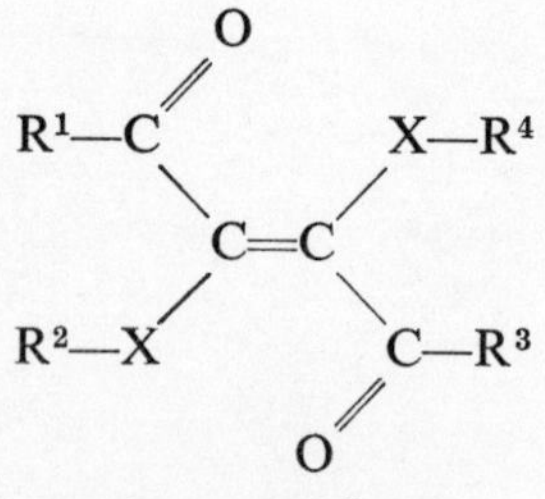

$$R^1-C(=O)-C(R^2-X)=C(X-R^4)-C(=O)-R^3$$

cis- or trans-form
X = electron donor
(10.2)

Carbonyl Dyes

According to the HMO and PP calculations carried out by M. Klessinger and W. Lüttke[1] (cf. section 2.1), the ringfree group (**10.2**) has all the typical characteristics of indigo; it is, therefore, the source of the colour. Klessinger and Lüttke's precepts appear to be substantiated by the fact that the compounds they obtained confirmed their predictions (cf. Table 10.1). Nevertheless, with all technically important indigoid dyes, R^1 and R^2 (or R^3 and R^4) are parts of an aromatic ring system. There are several reasons for this.

1. The synthesis of indigo dyes is simple when R^1 and R^2 (or R^3 and R^4) are components of an aromatic ring (cf. section 10.1.2). So far, it has not been possible to produce the simplest indigo structure ($R^1 = R^2 = R^3 = R^4 = H$; $X = NH$), that is to say, the parent substance of indigo. However, the compound (**10.3**) ($X = NH$) synthesized by Lüttke and coworkers comes close to that structure.
2. Conjugation of the colour-bearing structural elements with aromatic systems endows these dyes with the necessary affinity for the substrate. This affinity forms the basis of the application of indigo derivatives as vat dyes (cf. section 10.5.5.).
3. In comparison with the previously calculated values, the absorption maxima of the ringless compounds (**10.7**) and (**10.8**) (cf. Table 10.1) exhibit a marked hypsochromic displacement. This deviation is caused by steric interactions of the substituents R^1 and R^2 (or R^3 and R^4) which disturb the coplanarity of the compounds. The introduction of a covalent bond ((**10.3**)–(**10.6**)) forces the system into a planar configuration, thus preventing the hypsochromic shift.

It is a characteristic of indigoid dyes that an extension of the central double bond system by one double bond produces not a bathochromic displacement (cf. Chapter 6) as with cyanines, merocyanines or polyenes, but a hypsochromic shift of the bands.

The chemical structures of the most important indigoid dyes can be classified in accordance with either the electron donors X (cf. Table 10.2) or considerations of symmetry.

Indigoid compounds which either have only one electron donor X (semi-indigo derivatives, for example indole-aryl- and thionaphthene-aryl-indigos) or different electron donors X in the 1- and the 1'-positions (mixed indigo derivatives, for example indole-thionaphthene-indigos) are classified as 'unsymmetrical indigoids'. The large number of structural variations is markedly increased by the possibility of 2,3'-linkages (indirubin derivatives) and 3,3'-linkages (isoindigo derivatives).

Indigo (2,2-bisindole-indigo) (**10.12**), after which this group of carbonyl dyes is named, is one of the oldest known dyes. It is a derivative of the colourless indican, and it occurs principally in the indigo plant (*Indigofera tinctoria*; India, Java and China) and dyers' woad (*Isatis tinctoria*; Europe).

H. Cassebaum[2] has described the history of indigo. A. von Baeyer elucidated its structure in 1870–1883 and K. Heumann succeeded in establishing the first technically practicable synthesis in 1890 (cf. section 10.1.2).

Table 10.1 Comparison between calculated and observed absorption bands of indigo compounds[1]

Compound:

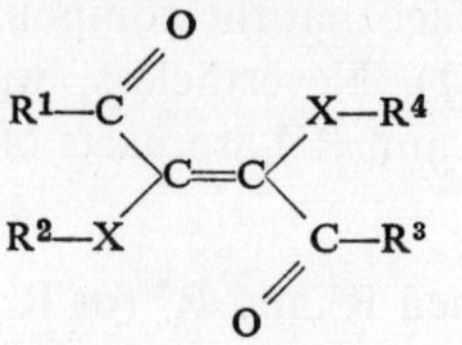

$$R^1-C(=O)-C(X-R^2)=C(X-R^4)-C(-R^3)=O$$

Compound no.	in cis- or trans-form	Calculated λ_{max}(nm)	Observed λ_{max}(nm) in cyclohexane	log ε
(10.3)	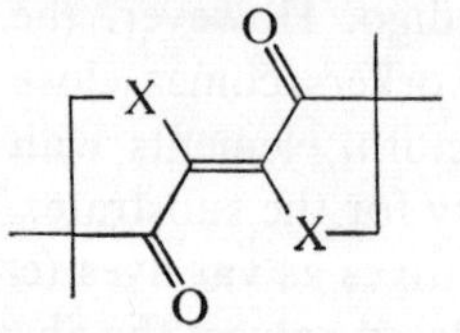	453 (X = NH)	450 (X = S) 484 (X = Se) 480 (X = NH)	4·13
(10.4)	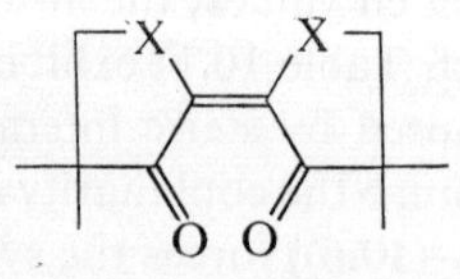	398 (X = NH)	394 (X = S)	4·05
(10.5) (R = H) (10.6) (R = CH$_3$)	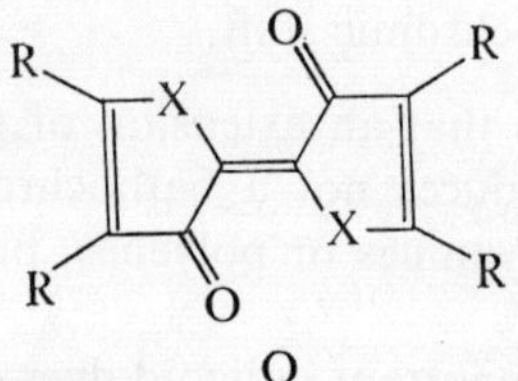	486 (X = NH)	502 (X = S) 528 (X = NH)†	3·85†
(10.7)	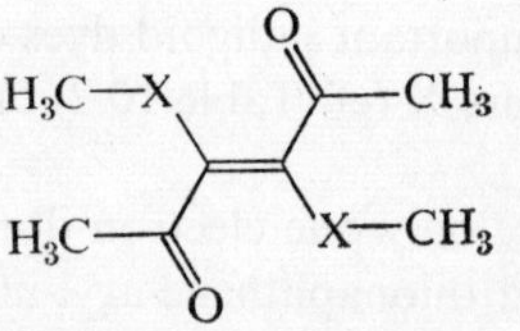	453 (X = NH)	315 (X = S)	3·95
(10.8)	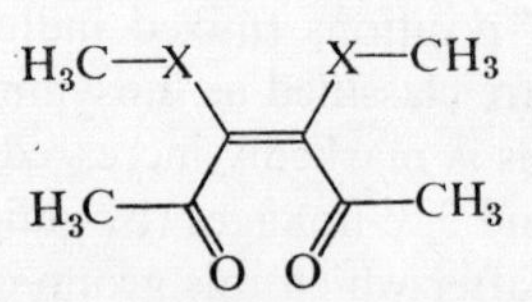	398 (X = NH)	318 (X = S)	3·94
(10.9) (10.10) (10.11)	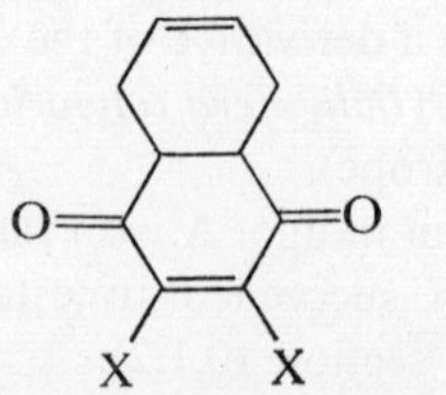	415 (X = NH$_2$)	434 (X = NHC$_6$H$_5$) 374 (X = SC$_6$H$_5$) 386 (X = SCH$_3$)	4·02 3·72 3·83

† The spectrum of compound (10.6) was measured in chloroform.

Table 10.2 Classification of indigo dyes according to electron donors

X	Name	λ_{max}(nm)(ethanol; 10 mg/l)	log ε
NH	Indigo (10.12)	606	4·23
Se	Selenindigo (10.13)	562	4–5
S	Thioindigo (10.14)	543	4–5
O	Oxindigo (10.15)	432	4–5

Indigo crystallizes (from high boiling solvents such as nitrobenzene and aniline) in the form of prisms with a copper-red lustre. It is almost insoluble in dilute acids and in aqueous alkaline solutions, but it dissolves in concentrated sulphuric acid. It is sparingly soluble in organic solvents.

X-ray structural analysis and spectroscopic measurements have established that indigo has a *trans*-configuration both in the crystalline state and in solution. In contrast to thio-, ox- and selenium-indigo, and N-substituted indigo derivatives, it is not possible to enrich the *cis*-form of indigo even photochemically, since the *trans*-form of the NH $\cdots$ O=C hydrogen bonds is energetically highly favourable.[3]

The unusually high melting point (390–392°C) and the poor solubility of indigo can be explained in terms of its crystal structure; the X-ray diagram[4] shows that indigo in the solid state forms a hydrogen-bonded polymer, in which each indigo molecule is linked to four surrounding molecules.

In the concentration range 10^{-5}–10^{-6} mol/litre indigo dissolves in non-polar solvents; the solutions are red to red/violet in colour and the indigo is present mainly as monomer, whilst in polar solvents it is associated and the solutions are blue. The different colours are caused by solvatochromy (section 2.1) and association, the latter being a function of the polarity of the solvent (cf. Table 10.3).

In the preparation of substituted (thio-)indigo dyes, the desired substituents are introduced either before ring closure to form the (thio-)indoxyl derivative, or by direct substitution of the (thio-)indigo compounds. Apart from the possible disturbing influence of the ring closure reaction, the first method permits the introduction of substituents into any desired position. In contrast, with direct substitution, substituents can only be brought into the appropriate position in a predetermined sequence in accordance with the rules of electrophilic aromatic substitution. With the introduction of electron acceptors (sulphonation, nitration) the 5,5'-position is first occupied (X: *p*-directing; >C=O: *m*-directing). Subsequently the 7,7'-position can also be substituted (X: *o*-directing; >C=O and 5-NO_2: *m*-directing). Electron donors also first enter the 5,5'-position and then the 7,7'-position.

Table 10.3 Influence of solvent on absorption†

Compound	Solvent	Dielectric constant (20°C)	λ_{max} (nm)
Indigo	Vapour	—	540
	Carbon tetrachloride	2·238	588
	Xylene	2·3–2·6	591
	Ethanol	24·30	606
	Dimethylsulphoxide	46·35 (25°C)	620
	Solid, in KBr	—	660
N,N'-dimethyl-indigo	Carbon tetrachloride	2·238	640
	Benzene	2·284	644
	Chloroform	4·806	653
	Ethanol	24·30	656
	Ethanol/H$_2$O = 1:2	24·3–80·4	672
	Solid, in KBr	—	672

† Owing to the poor solubility of the compounds the ε-values obtained are very inaccurate.

The o/p-directing influence of electron donors, however, makes substitution at the 4,4'-position possible.

The effect of substituents on the absorption maximum can be predicted with the aid of certain colour rules (distribution rules for auxochromes[5]) and they can be defined quantum-chemically by using the MIM method[6] (cf. section 2.1). Since (thio-)indigo components have already one electron donor in the o-position, the largest bathochromic shift can be expected when a further electron donor is introduced at the 5-position (2,5-position in respect of the electron acceptor). In fact, it is seen that there is a displacement of the absorption band with the longest wavelength by the introduction of electron donors, resulting in the following sequence of positions:

6- <7- <4- <5- (cf. Table 10.4).

The absorption band of the non-substituted compound generally occupies an intermediate position. The effect of electron acceptors is analogous. Thus, the 6-nitro derivative exhibits a larger bathochromic shift than the 5-nitro derivative. Halogen (mainly) and some methyl, methoxy and benzo derivatives are technically important as dyes.

Table 10.4 Effect of substituents on λ_{max}

No.	Compound Name	λ_{max}(nm) Ethanol	Tetrachlorethane
(10.16)	6.6'-dichlorindigo	570	590
(10.17)	7.7'-dichlorindigo	590	606
(10.12)	Indigo	606	620
(10.18)	4.4'-dichlorindigo	612	616
(10.19)	5.5'-dichlorindigo	615	620

10.1.2. Methods of synthesis used in indigo chemistry

In the first half of the 20th century, the application of numerous symmetrical and unsymmetrical indigo and thio-indigo compounds as dyes was investigated. Nowadays, at least thirty synthetic methods for producing indigo dyes are known.

This Chapter deals mainly with fundamental aspects of chemistry that are indispensable for the understanding of technical syntheses. In this connection, still unsolved scientific problems must be mentioned.

Nearly all syntheses occur, basically, in two separate steps which are dealt with in sections I and II. In the third section some of the general aspects of the reaction mechanisms of the individual steps are discussed.

I. Linkage of five-membered ring to a benzene or naphthalene derivative, thus forming (thio-)indoxyl (**10.22**), or a (thio-)isatin-2-anil (**10.24**).

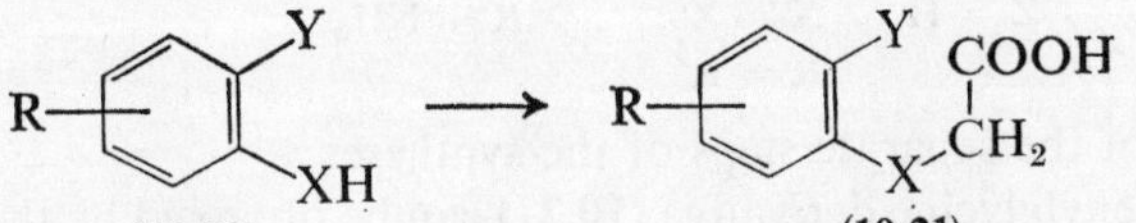

$$(10.20) \longrightarrow (10.21) \tag{10.1}$$

$X = $ NH or S
$Y = $ H, COOH or CN

Phenylglycine derivative ($X = $ NH)
Phenylthioglycol derivative ($X = $ S)

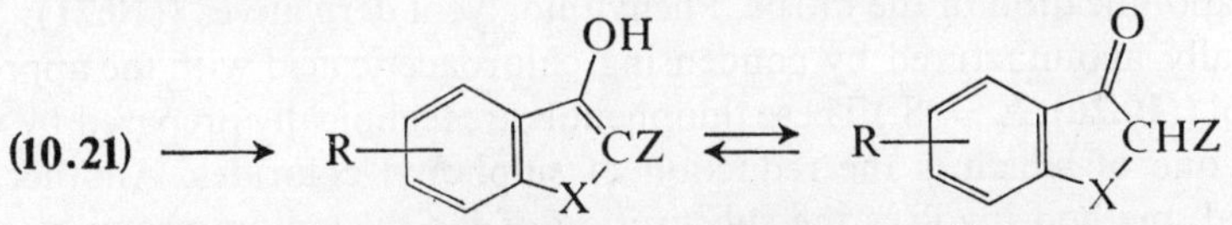

$$(10.21) \longrightarrow \text{Enrol-form} \rightleftharpoons (10.22) \text{ Keto-form}$$

where $\begin{array}{l} Y = \text{COOH,CN} \\ Y = H \end{array} \rightarrow \begin{array}{l} Z = \text{COOH} \\ Z = H \end{array}$

$$(10.22) \ Z = \text{COOH} \xrightarrow{H^\oplus} (10.22)\text{: } Z = H \tag{10.2}$$

Indoxyl ($X = $ NH)
Thioindoxyl ($X = $ S)

$(10.22) + $ (**10.23**) $R' = $ NO $\longrightarrow$
$Z = H$

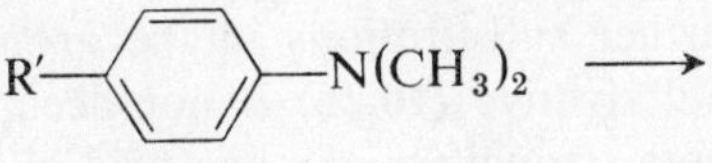

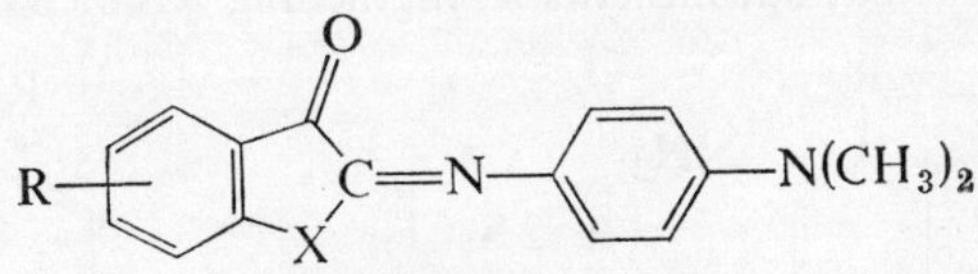

(**10.24**)

Isatin-2-anil ($X = $ NH)
Thioisatin-2-anil ($X = $ S) $\tag{10.3}$

II. Oxidative linkage of two equivalents of (thio-)indoxyl (**10.22**) for the production of symmetrical dyes ((10.4): $X = X'$, $R = R'$), or reaction of (thio-)indoxyl (**10.22**) with (thio-)isatin-2-anil (**10.24**†) for the synthesis of unsymmetrical derivatives ((10.5): $X \neq X'$, $R \neq R'$).

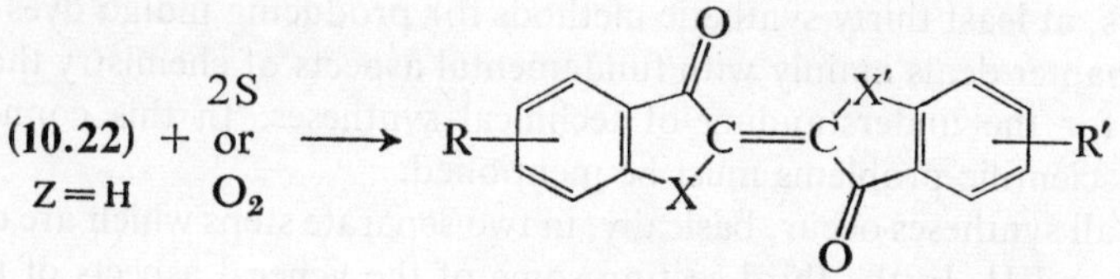

$$(\mathbf{10.22}) + \begin{array}{c} 2S \\ \text{or} \\ O_2 \end{array} \longrightarrow \text{(10.25)} \quad (10.4)$$

$$Z = H$$

$$R' = R$$
$$X' = X$$

$$(\mathbf{10.22}) + (\mathbf{10.24}) \longrightarrow (\mathbf{10.23}) + (\mathbf{10.25})$$
$$Z = H \quad X = X' \quad R = NH_2 \quad (10.5)$$
$$R = R'$$

III. Discussion of the separate steps of the syntheses:

(10.1): The phenylglycine derivative (**10.21**) can be obtained by the condensation of chloracetic acid with the aniline derivative. Nowadays, however, in the synthesis of indigo, it is prepared from aniline, formaldehyde and sodium cyanide with subsequent saponification of the nitrile. Phenylthioglycol derivatives ((**10.21**): $X = S$) are generally manufactured by condensing chloroacetic acid with the appropriate thiophenol ((**10.20**): $X = S$.) These thiophenols are technically prepared by various methods, one of which is the reduction of sulphonyl chlorides. Another, much more usual, method involves the substitution of the diazonium group by sulphur compounds such as potassium ethyl xanthate or sodium disulphide. Thio-phenol is obtained by hydrolysis of the primary xanthate compound in the first case, and by reduction of the disulphide compound in the second case. The Herz reaction is, industrially as well as scientifically, a very interesting method of introducing aromatic mercapto groups. In this case, an aromatic amine is converted into the corresponding o-aminothiophenol (10.6), (10.7) by reaction with sulphur chloride. This method frequently involves further substitutions in the aromatic nucleus. When the p-position of the aniline derivative (**10.26**) is not occupied ($R = H$) chlorination occurs. When substituents strongly conferring acid character (NO_2, CN, COOH) are present in the o- or p-position these, too, are replaced by chlorine, but substituents strengthening basic character such as methyl, hydroxy and

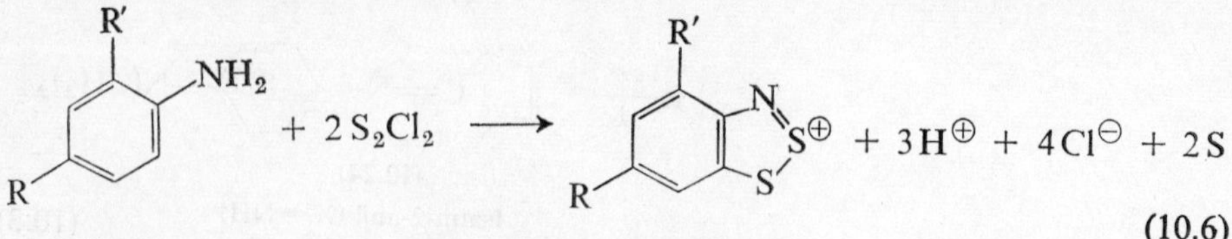

$$(\mathbf{10.26}) + 2\,S_2Cl_2 \longrightarrow (\mathbf{10.27}) + 3\,H^{\oplus} + 4\,Cl^{\ominus} + 2\,S \quad (10.6)$$

† Instead of the anil, isatin chloride, which can be made from isatin, may be used.

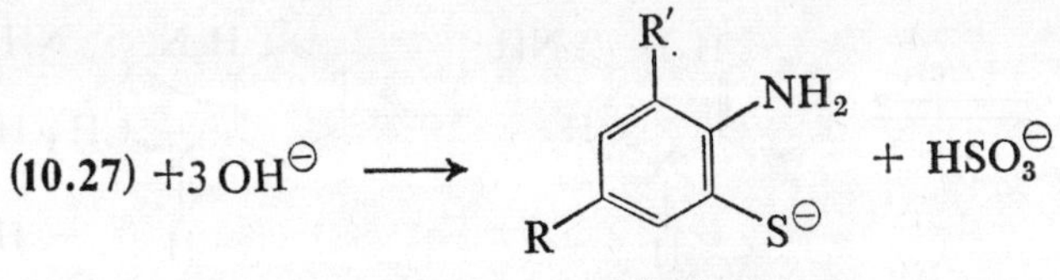

Carbonyl Dyes

$$(10.27) + 3\,OH^\ominus \longrightarrow \underset{(10.28)}{\text{[aromatic ring with } R', NH_2, R, S^\ominus\text{]}} + HSO_3^\ominus \qquad (10.7)$$

amino groups are not replaced. From experimental results it may be conjectured that electrophilic as well as nucleophilic substitutions can occur in the course of the Herz reaction. The mechanism of this reaction is still under discussion.[7]

(10.2): In Heumann's classical synthesis (1890) phenylglycine-*o*-carboxylic acid ((**10.21**): $X = NH$, $Y = COOH$) is converted by fusion with NaOH at 200–220°C, with exclusion of air, into indoxyl-2-carboxylic acid ((**10.22**): $Z = COOH$) which can be oxidized by air to indigo (**10.12**). This ring closure is assumed to proceed in an analogous manner to the Dieckmann condensation. For this reason it can also be performed with appropriate phenylthioglycol-*o*-carboxylic acids ((**10.21**): $X = S$, $Y = COOH$) or -*o*-cyanides (10.8).

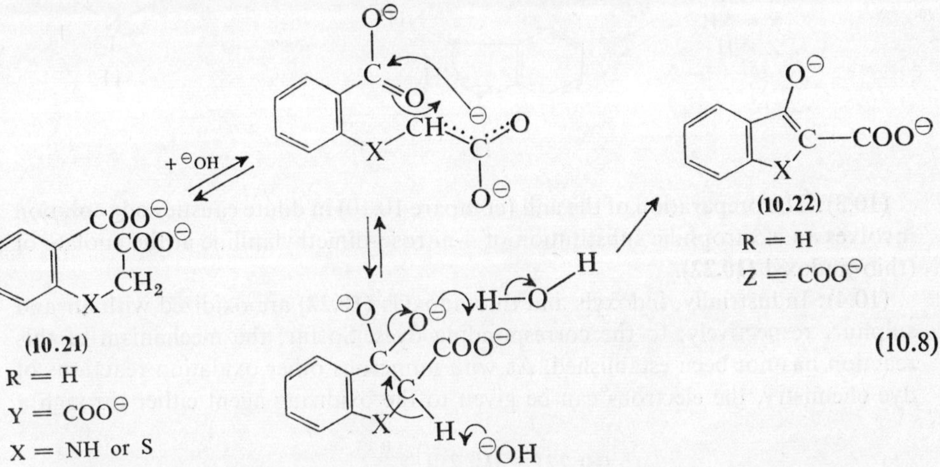

$$(10.8)$$

(**10.21**) $R = H$, $Y = COO^\ominus$, $X = NH$ or S

(**10.22**) $R = H$, $Z = COO^\ominus$

Pfleger's synthesis (1901) is much more efficient; the alkaline melt is carried out in the presence of sodamide (10.9). Here, the phenylglycine itself ((**10.21**): $X = NH$, $Y = H$) can be converted directly into indoxyl (**10.22**) (10.9). However, this method fails with phenylthioglycol and N-alkylated glycine derivatives that do not have a carboxyl or cyano group in the *o*-position. In such cases, acid cyclization is carried out, either directly with chlor-sulphonic acid or via the acid chloride in the presence of anhydrous AlCl$_3$. The preparation of an acid chloride involves a nucleophilic substitution of a hydroxyl ion by a chloride ion at the carbonyl carbon atom, whilst a ring closure involves an electrophilic substitution of a ring proton by an acylium cation (Friedel–Crafts reaction).[8]

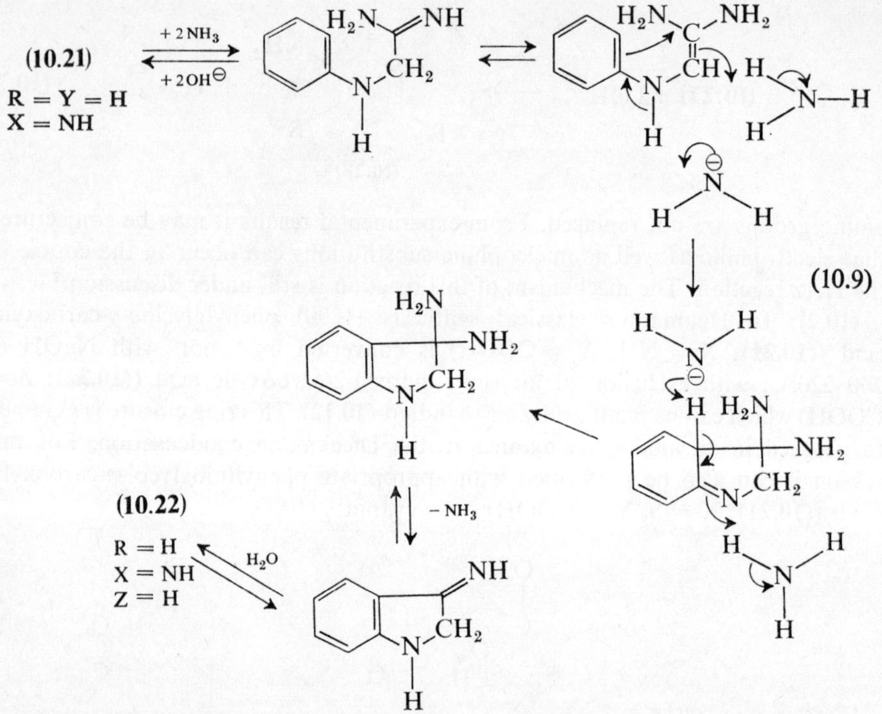

(10.3): The preparation of the anil (compare 10.10) in dilute caustic soda solution involves an electrophilic substitution of 4-nitroso-dimethylaniline at the enolate of (thio-)indoxyl (**10.22**).

(10.4): Industrially, indoxyls and thioindoxyls (**10.22**) are oxidized with air and sulphur, respectively, to the corresponding dyes. So far, the mechanism of this reaction has not been established. As with numerous other oxidation reactions of dye chemistry, the electrons can be given to the oxidizing agent either through a

$$(\mathbf{10.22}) + (\mathbf{10.23}) \underset{}{\overset{B}{\rightleftarrows}}$$
$$Z = H \qquad R = NO$$

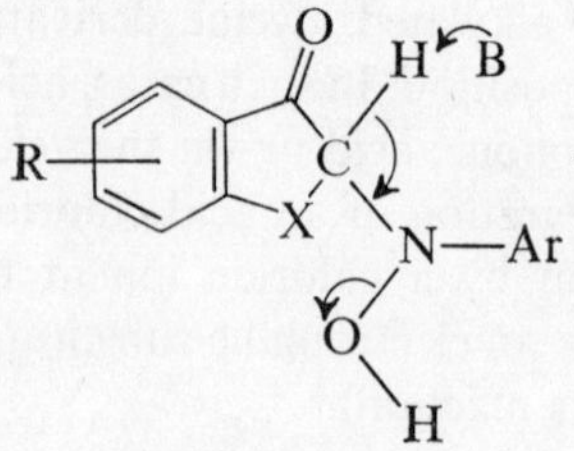

$$\longrightarrow (\mathbf{10.24}) + HB^{\oplus} + OH^{\ominus} \qquad (10.10)$$

hydride ion transfer mechanism or by a stepwise process (cf. section 3.5). The following applies to the dimerization mechanism (S = oxidising agent):
First possibility: A hydride ion transfer mechanism (10.11).

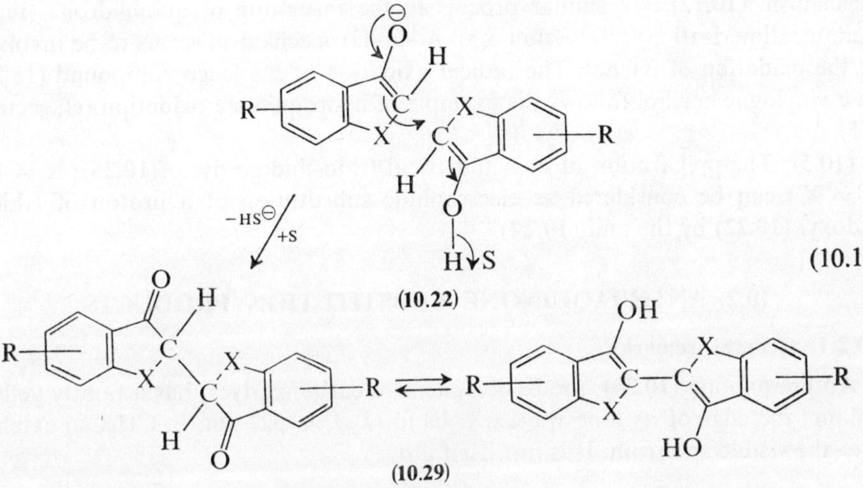

(10.11)

(10.22)

(10.29)

Second possibility: A radical mechanism (10.12).

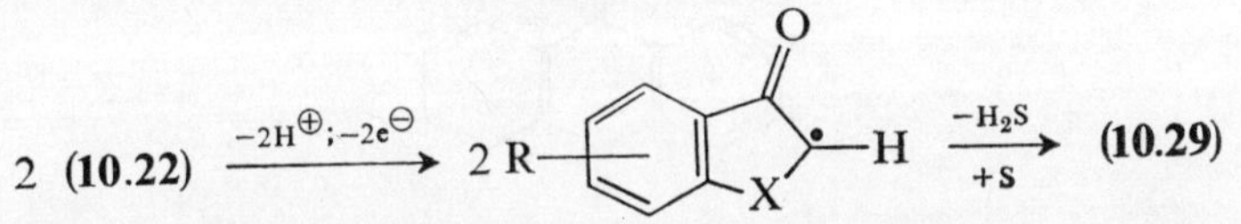

$$2\ (\mathbf{10.22}) \xrightarrow{-2H^{\oplus};\,-2e^{\ominus}} 2\ R\text{—[...]}\text{—H} \xrightarrow[+S]{-H_2S} (\mathbf{10.29}) \qquad (10.12)$$

The oxidation mechanism of the leuco compound (**10.29**), to (thio-)indigo, is analogous (S = oxidising agent):
First possibility: A hydride transfer mechanism (10.13).

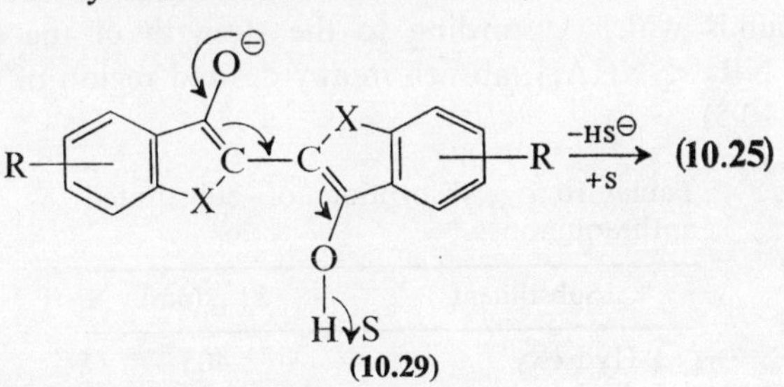

$$\text{[...]} \xrightarrow[+S]{-HS^{\ominus}} (\mathbf{10.25}) \qquad (10.13)$$

(10.29)

Second possibility: A radical mechanism (10.14).

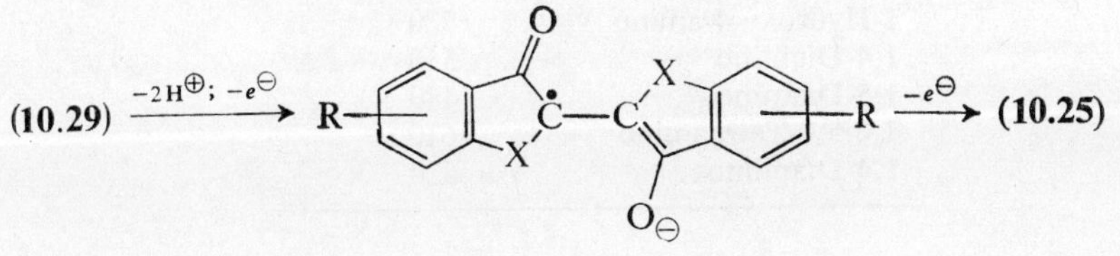

$$(\mathbf{10.29}) \xrightarrow{-2H^{\oplus};\,-e^{\ominus}} R\text{—[...]}\text{—R} \xrightarrow{-e^{\ominus}} (\mathbf{10.25}) \qquad (10.14)$$

Oxygen, and also the resulting indigo (**10.25**), could serve as hydride acceptors in the dimerization reaction (10.11). The latter would give the leuco form (**10.29**), which is then re-oxidized with atmospheric oxygen. Dimerization by the radical mechanism (10.12) is a similar process to the formation of quinhydrone (semi-quinone dimerization) (cf. section 3.5). A similar mechanism seems to be involved in the oxidation of phenol. The radical oxidation of the leuco compound (**10.29**) to a vinylogue enediol follows the example of hydroquinone oxidation (cf. section 3.5).

(10.5): The preparation of unsymmetrical (thio-)indigo dyes ((**10.25**): R ≠ R', X ≠ X') can be considered as electrophilic substitution of a proton of (thio)-indoxyl (**10.22**) by the anil (**10.24**).

10.2. ANTHRAQUINONE SUBSTITUTION PRODUCTS

10.2.1. General remarks

Anthraquinone (**10.30**), the basic system of carbonyl dyes, has a faintly yellow colour; the edge of its long wave u.v. band (λ_{max} = 327 nm, in CH_2Cl_2) extends into the visible spectrum. It is not itself a dye.

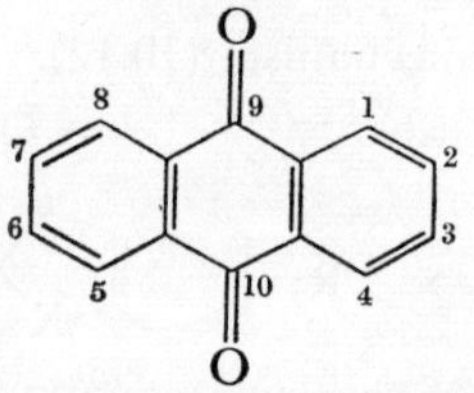

(**10.30**)

The introduction of relatively simple electron donors already gives anthraquinone compounds which, according to the strength of the electron donors ($OH < NH_2 < NR_2 < NHAr$), absorb in any desired region of the visible spectrum (cf. Table 10.5).

Table 10.5 Absorption of substituted anthraquinones.[9]

Substituent	λ_{max}(nm)
1-Hydroxy	405
1-Amino	465
1-Methylamino	508
1-Hydroxy-4-amino	520
1,4-Diamino	550
1,5-Diamino	480
1,4,5,8-Tetraamino	610
1,4-Dianilino	620

Here, substituents in the α-positions (e.g. 1-, 1,4- or 1,5-) cause the largest bathochromic shifts; this can be determined with the aid of the MIM method (cf. section 2.1: distribution rule of auxochromes). H. Labhart[9] has compared the experimentally-determined optical energy transitions of simple substituted anthraquinone derivatives with the LCAO-MO data calculated by A. Kuboyama.[10] These simple substituted compounds are frequently used directly as dyes (cf. section 10.5) or they serve as intermediate products for the manufacture of coloured polycondensed carbonyl compounds (e.g. indanthrone (10.47)). The position of the substituents influences, not only the absorption maximum, but also some of the other properties, as, for example, the resistance to sublimation of disperse dyes (section 10.5.2), affinity for the substrate, light fastness and vatting properties (section 10.5.5) owing to the formation of hydrogen bonds between substituent and the carbonyl group.

Anthraquinone derivatives that have hydroxyl or amino groups in the β-position, and are thus capable of forming intermolecular hydrogen bonds, generally exhibit better resistance to sublimation, better solubility and better affinity for textile substrates, than α-substituted compounds. On the other hand, intramolecular hydrogen bonds reduce the acidity of the carbonyl groups in the peri-position (more positive redox potential) which, generally, is advantageous in respect of wash and light fastness. X-ray investigations have shown that the relative strengths of intermolecular and of intramolecular hydrogen bonds with different hydroxy- and aminoanthraquinones are reflected in their melting points;[11] intermolecular hydrogen bonds can raise the melting point by up to about 100°C. Analogous relations between chemical structure and gas chromatographic retention times supply information on the formation of intermolecular hydrogen bonds between substrates and simple anthraquinone dyes.[25]

The history of anthraquinone chemistry is characterized by a vigorous development at the turn of the century. C. Graebe and A. Liebermann (1868) were responsible for the development of anthraquinones, which resulted in the elucidation of the structure of alizarin (10.34). The following 50 years had a decisive influence on organic chemistry, owing to the progress of anthraquinone chemistry. Names such as W. H. Perkin, H. Caro, R. E. Schmidt, R. Bohn, F. Ullmann, R. Scholl, O. Bally and M. Kunz, to mention just a few, are inseparably connected with this development.

10.2.2. Structures and principles of preparation

There are many methods of synthesis in anthraquinone chemistry; their discussion in this book is, therefore, by no means exhaustive. Although optimum experimental conditions for carrying out most reactions at the anthraquinone nucleus were already known fifty years ago, the exact course of these reactions is largely unknown; it is partly attributable to the fact that, on the one hand, identification of the by-products, as well as the reaction mechanism under widely different reaction conditions (alkali melts, high pressure and temperature) is very difficult, and, on the other hand, because there are preexisting acid, base and redox equilibria before the substitutions proper. Nevertheless, in order to arrive at a certain system and an understanding of the reaction mechanisms of the most important technical syntheses, the authors will discuss in this Chapter the principles underlying the preparation of anthraquinone dyes, with examples of some characteristic reactions, in accordance with the explanations given in Chapter 3.

There are three different ways of producing hydroxyanthraquinones: by direct synthesis of substituted compounds beginning with phthalic anhydride and the corresponding substituted benzene derivatives (Friedel–Crafts reaction and ring closure of the resulting 2-benzoylbenzoic acid derivatives using sulphuric acid), by substitution of hydrogen (oxidation), and by the substitution of nucleofugal leaving groups, e.g. $SO_3^\ominus$-, halogen- and NO_2-groups. The last two methods can be illustrated by means of the classical example of the synthesis of alizarin (**10.34**) and its derivatives. Starting with anthraquinone-2-sulphonic acid ((**10.31**): $X = SO_3^\ominus$), alizarin (**10.34**) is obtained by fusion with potassium hydroxide, hydrogen (10.15) and a sulphonic group ((10.16): $X = SO_3^\ominus$) being replaced by OH. The position in which substitution first occurs is uncertain, but, from analogous reactions, it is conjectured that the hydroxyl group first enters the 1-position and forms the hydroquinone compound (**10.32**). Starting with 2-nitroanthraquinone ((**10.31**): $X = NO_2$), it is possible to detect 1-hydroxy-2-nitroanthrahydroquinone ((**10.32**): $X = NO_2$) as an intermediate product, in addition to alizarin. As equations (10.15) and (10.16) show, the question of the sequence of the substitution can be of decisive importance for the isolation of intermediate products. If hydrogen is first substituted (10.15), preoxidation is necessary for the subsequent exchange

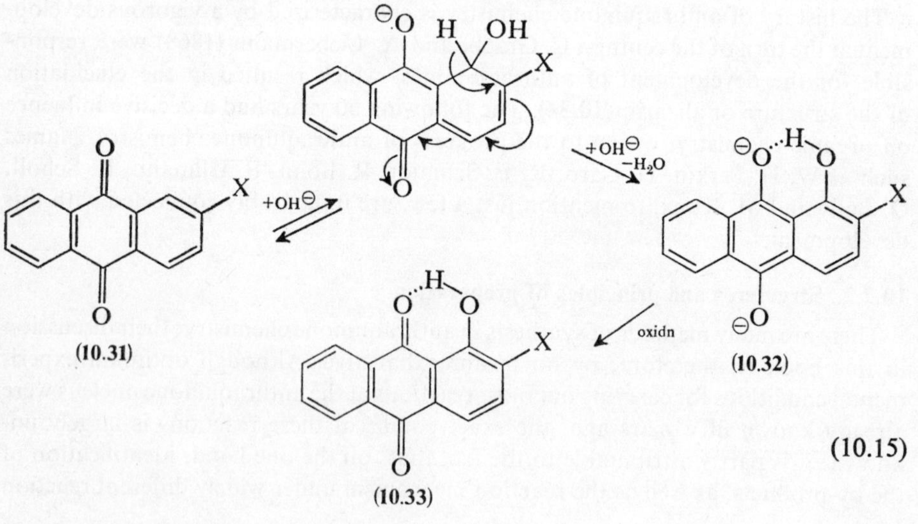

(10.15)

of X (10.16); when the sequence is reversed, oxidation is the last reaction step. Apart from a sulphonic, nitro or halogen group, hydrogen (= anthraquinone: (10.30)) can be used as substituent X. In this case no hydride ion is split off, but, by analogy with reaction (10.15), it gives rise to the corresponding hydroquinone derivative by proton release, and is subsequently oxidized to the final product with the loss of 2 electrons. With X in the 1-position, it is possible to produce 1-hydroxy derivatives with calcium hydroxide, and 1,2-hydroxy derivatives with potassium hydroxide.

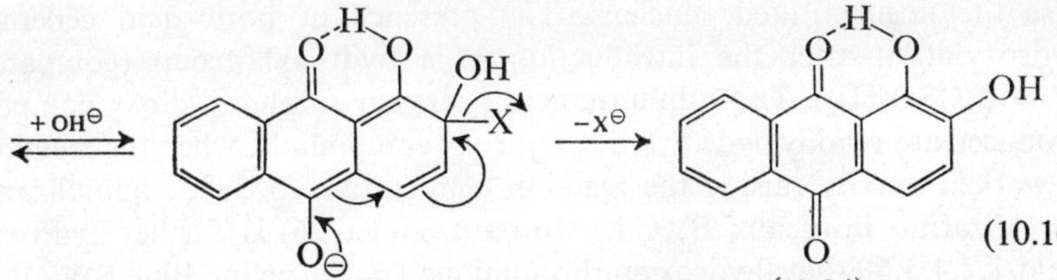

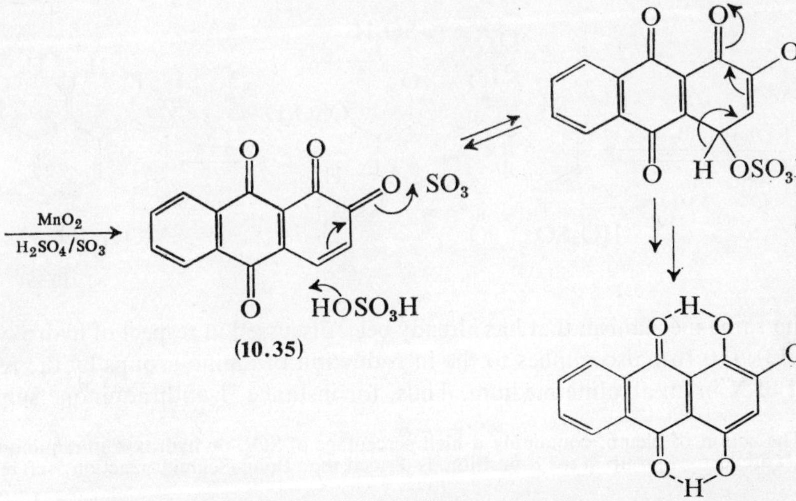

The hydrogen atoms in the anthraquinone nucleus can be replaced by hydroxyl groups not only in alkaline but also in acid media. The choice of the oxidant determines the substitution site. As will be seen from (10.17)–(10.19), it is of decisive importance whether the oxidizing agent can react directly with the educt (10.34) or only with the substitution product (cf. (10.18)). Strong oxidizing agents (MnO_2, PbO_2) can oxidize alizarin (10.34) into 1,2,9,10-anthradiquinone (10.35) and this facilitates the addition of a nucleophilic particle at the substituted ring (cf. (10.17)). In contrast, oleum exerts an oxidizing effect only after the nucleophilic addition of the sulphato group (10.18). The addition occurs at the more electrophilic, in this case the unsubstituted, nucleus. The presence of boric acid generally delays hydroxylation after the introduction of a hydroxyl group (compare (10.18): $M = B(OSO_3H)_2$). The sulphuric acid esters, or sulphuryl-dioxy compounds first produced are readily hydrolysed to hydroxy compounds, when the reaction mixture is worked up. By raising the reaction temperature to 38°C, quinalizarin **(10.39)** (= Alizarine Bordeaux B, C.I. Mordant Violet 26) is further hydroxylated† to yield 1,2,4,5,6,8-hexahydroxyanthraquinone (= Alizarine Blue SWR).

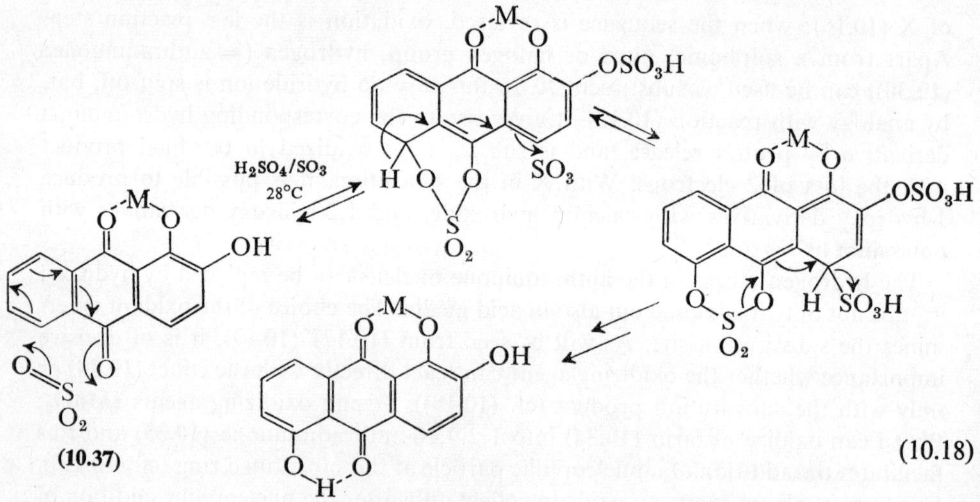

(10.38)
$M = B(OSO_3H)_2^-$ or H

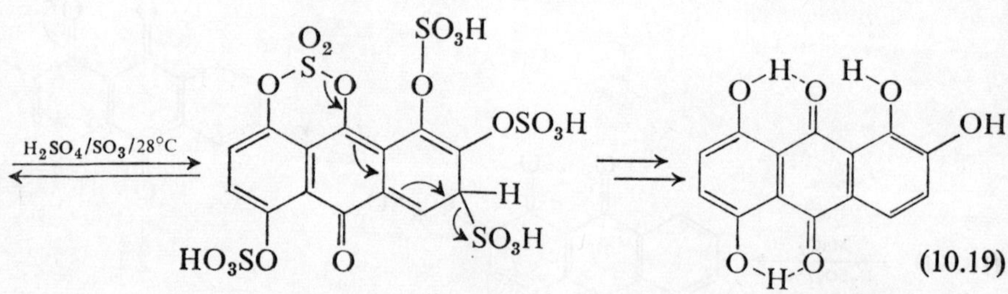

The same mechanism that has already been discussed in respect of hydroxylation ((10.15), (10.16)) also applies to the introduction of amino groups by the replacement of X in an alkaline medium. Thus, for instance, 1-anthraquinone sulphonic

† The action of oleum, containing a high percentage of SO_3, on hydroxyanthraquinone, with at least a hydroxyl group in the α-position, is termed the "Bohn–Schmidt reaction" (cf. reference 26).

acid in 30 per cent ammonia at 175°C can be transformed into 1-aminoanthraquinone, which, in more vigorous conditions (200°C) and in the presence of an oxidizing agent, is converted into 1,2-diaminoanthraquinone. 1-phenylaminoanthraquinone is readily converted into the 1,4-dianilino derivative at 60°C by sodium anilate and atmospheric oxygen. These principles of preparation are largely applied to the synthesis of 1,2-anellated dihydroazine derivatives: indanthrone (**10.47**) is produced either by the condensation of two molecules of 1-bromo-2-aminoanthraquinone (copper catalyzed nucleophilic substitution of halogen by amino group (cf. (10.25)) or by oxidative dimerization of 2-aminoanthraquinone (**10.20**)–(**10.22**). In equation (10.22), only two of the numerous possible reaction mechanisms are indicated, in which 2-amino-1,2'-anthraquinonylamine (**10.45**) or its reduced form (**10.43**) can be condensed to indanthrone (**10.47**). Guidelines in the field of indanthrone synthesis are the contributions of W. Bradley and coworkers.[27] The capacity of the compounds (**10.43**) and (**10.45**) to tautomerize makes it possible for them to cyclize in the sterically-preferred equilibrium form. Consequently, when NH is replaced by NCH_3, ring closure becomes more difficult. Because of the multiplicity of the preequilibria and the resulting reaction possibilities (cf. (10.20)–(10.23)), it is extremely important to employ optimum reaction conditions when producing only one particular reaction product; relatively small changes of temperature and concentration may give rise to entirely different products. For instance, below 220°C an alkaline melt of 2-aminoanthraquinone (**10.40**) yields considerable quantities of alizarin (**10.34**), but above 220°C flavanthrone ((**10.76**): X = N; cf. (10.23)), among other products, is produced. In order to produce mixed aminohydroxyanthraquinones, amino groups may be introduced indirectly by nitration and subsequent acid ($SO_3/H_3BO_3/S$) or alkaline (Na_2S) reduction.

The preparation of 1,4-diaminoanthraquinone compounds, and their alkyl and aryl derivatives is of great industrial importance. The corresponding 1,4-dihydroxyanthraquinone (**10.49**) is first reduced to the leuco compound (**10.50**); this behaves as a vinylogue acid derivative and reacts stepwise with aliphatic and aromatic amines (10.24).

The most important industrial method of introducing amino groups into the anthraquinone nucleus involves the nucleophilic substitution of halogen. This reaction can be catalysed by metallic copper and by cupric or, rarely, cuprous ions. All copper-catalysed halogen substitutions are known as 'Ullmann reactions'. The reactivity of halogen compounds increases in the order RCl < RBr < RI. The mechanism of this catalysis is not known; if, despite this fact, the participation of copper in the substitution reaction is formulated (cf. (10.25), (10.27) and (10.28)), it is done only to facilitate the understanding of the complex relations between the numerous preequilibria, the varying reaction conditions and possible side reactions. Equation (10.25) shows that the bromoanthraquinone derivative (**10.53**), as a vinylogue acid bromide, forms an intermediate product with the nucleophilic

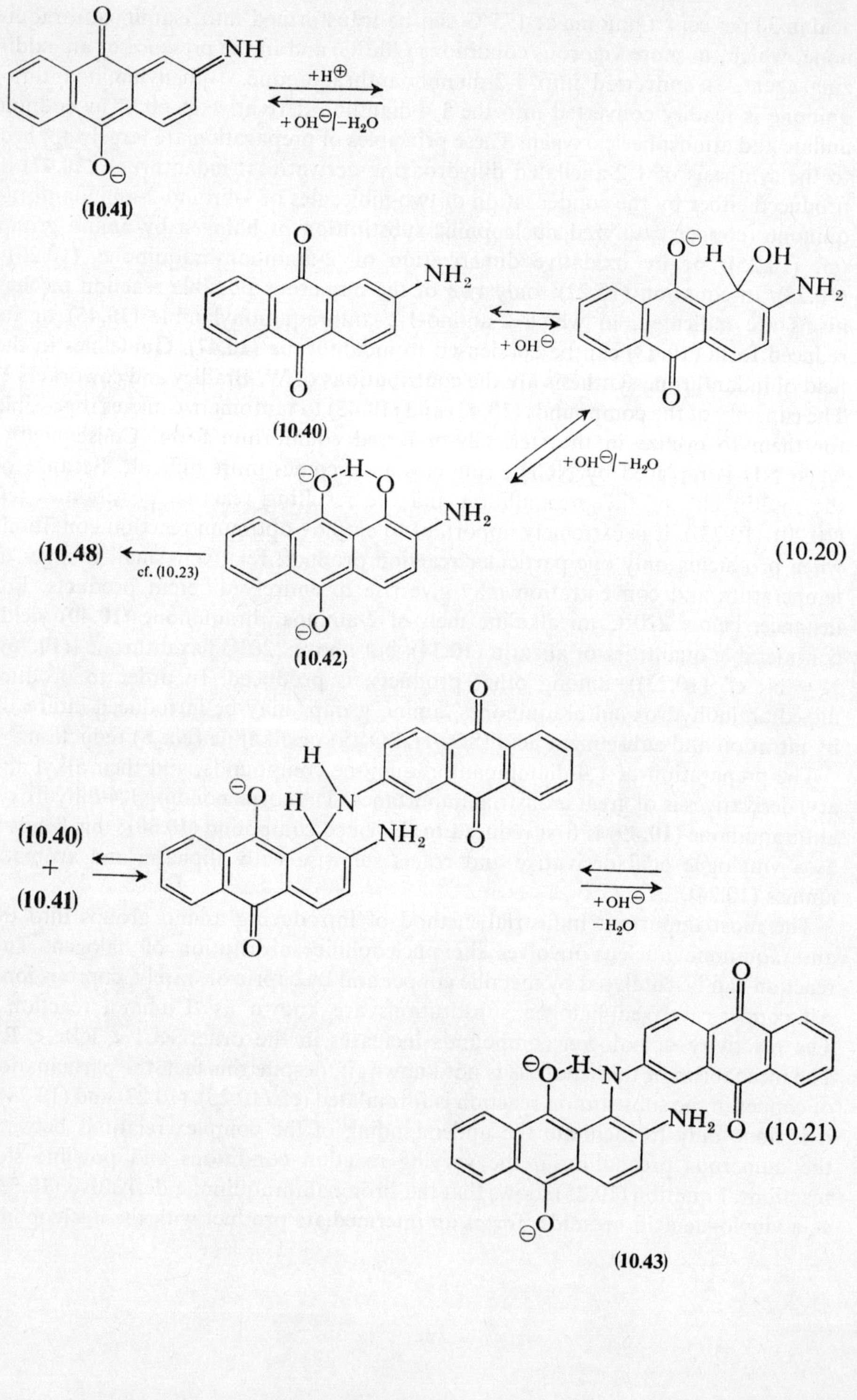

(10.41) ⇌ (+H⁺ / +OH⁻,−H₂O) [iminoanthrone structure]

(10.40) ⇌ (+OH⁻) [hydroxide adduct] ⇌ (+OH⁻/−H₂O) (10.20)

(10.48) ← cf. (10.23) — (10.42)

(10.40) + (10.41) ⇌ [dimeric intermediate] ⇌ (+OH⁻, −H₂O) (10.21)

(10.43)

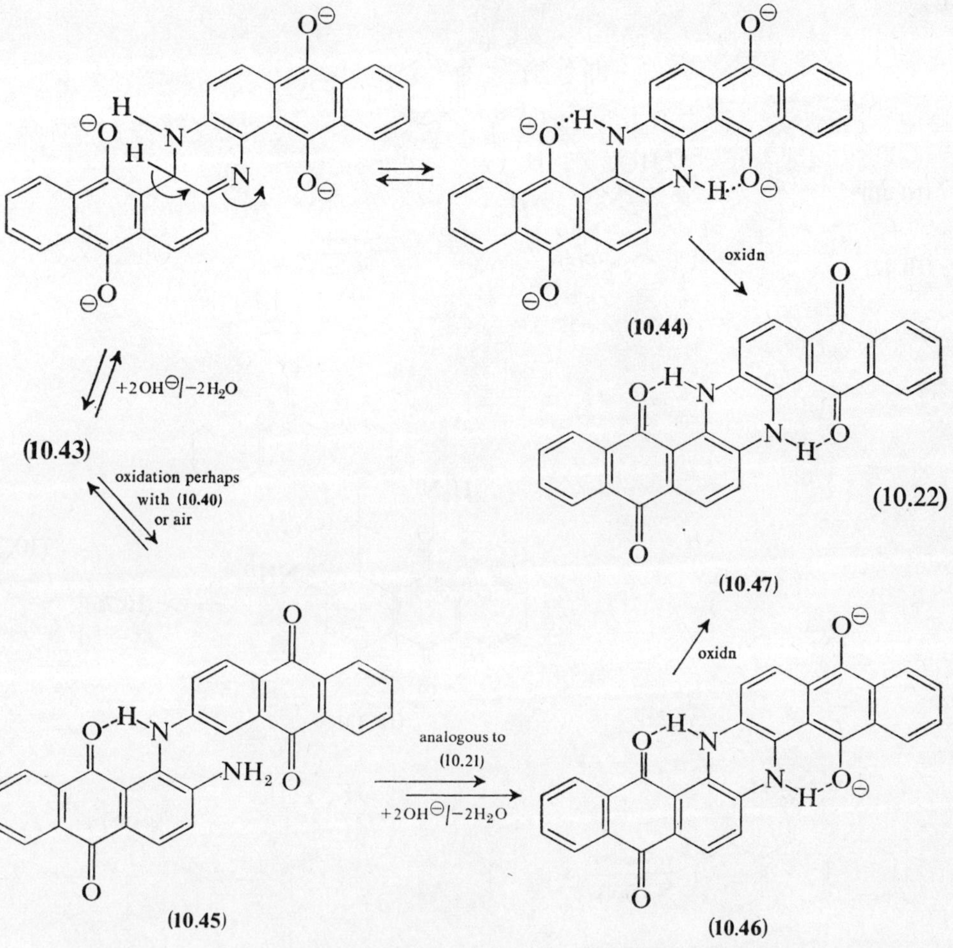

amine (**10.54**), which, in a rate-limiting, generally, base-catalysed step, splits off hydrogen bromide. The copper cation may exert its catalytic effect by the formation of an intermediate product (**10.55**), analogous to compound (**10.54**), at much higher concentrations thus accelerating the overall reaction. A mixture of cupric ions and copper bronze, in the presence of iodide, has the best catalytic effect. Metallic copper, as well as iodide ions, reduce the cupric ions resulting from oxidation to cuprous ions. One of the most important applications of this reaction is the preparation of acid anthraquinone dyes (cf. section 10.5.1) from bromoamino sulphonic acid ((**10.53**): $R^1 = H; X = SO_3^{\ominus}$). Apart from the expected 1-amino-4-hydroxy derivative (substitution of $Br^{\ominus}$ by $OH^{\ominus}$), compounds (**10.58**) and (**10.59**) occur as byproducts. These are frequently, but wrongly, attributed to a specific copper-catalysed side reaction. However, as equation (10.26) shows, these are

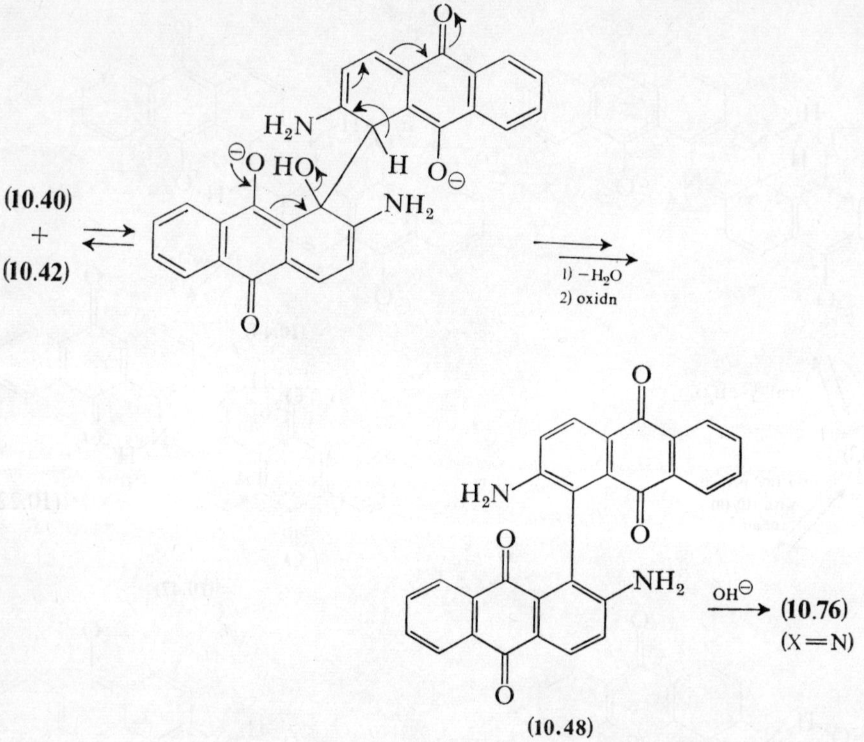

(10.40) + (10.42) ⇌ [intermediate] → 1) −H₂O 2) oxidn → (10.48)

(10.48) —OH⁻→ (10.76) (X=N)   (10.23)

(10.49) ⇌ reduction / oxidation ⇌ (10.50) ⇌ [+NH₂R¹] ⇌ [hydroxy amino intermediate]

[intermediate] —−H₂O→ (10.51) —+NH₂R² / oxidn→ (10.52)

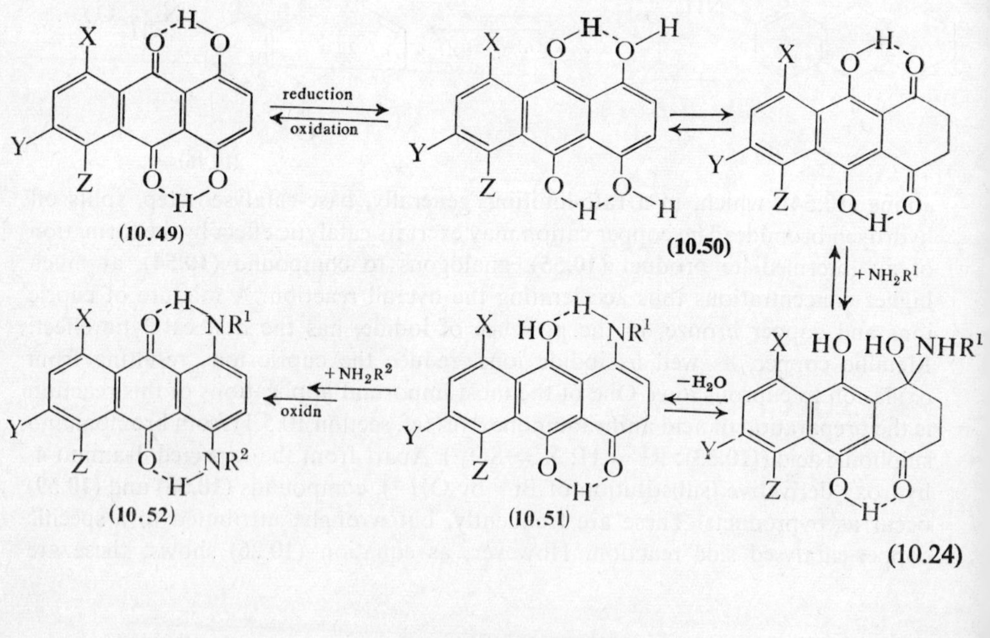

(10.24)

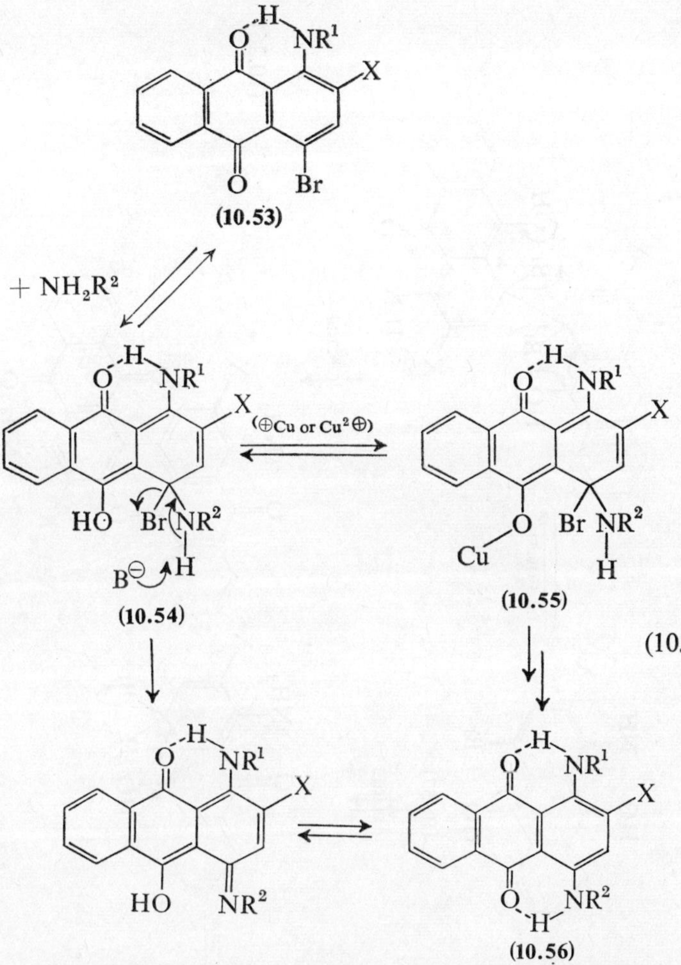

Carbonyl Dyes

(10.53)

+ NH$_2$R^2

(⊕Cu or Cu^2⊕)

(10.54) **(10.55)**

(10.25)

(10.56)

derived from the leuco compound **(10.57)** that may be produced by the reduction of **(10.53)**. Product **(10.56)** (as a *p*-phenylene diamine derivative) on the one hand, and the amine H$_2$NR^2, and sometimes copper (cf. (10.28)), on the other, can act as the reducing agent; thus, it was observed, for instance, that *p*-phenylene diamine (in contrast to the *m*-derivative), as the nucleophilic reagent, markedly favours the formation of **(10.58)**. Similarly, chloranthraquinones can be condensed with aminoanthraquinones to yield anthrimides **(10.69)**; in this case metallic copper or cupric ions largely act as catalysts. Therefore, it may be assumed that copper, as in C—C— condensations (10.27), participates as the nucleophilic reagent in the reaction. The reaction is generally carried out in nitrobenzene at 150–200°C. It is easier to prepare 1,1'- than 1,2'-anthrimides, but the latter are more readily produced than the 2,2'-derivatives.

(10.26)

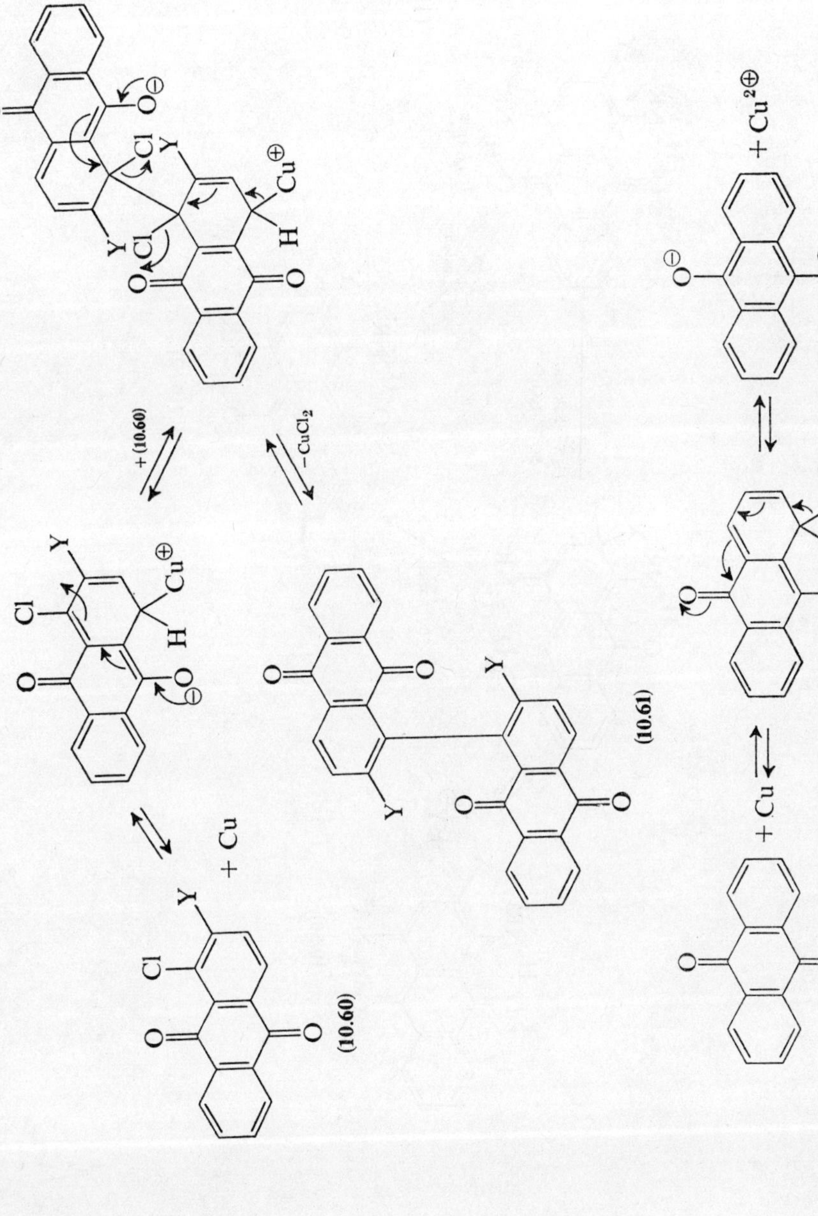

Carbonyl Dyes

(10.27)

(10.28)

(10.29)

Groups other than hydroxy and amino groups can be introduced into the anthraquinone nucleus by analogous methods. The preparation of 1-amino-2-alkoxy(or 2-aryloxy)anthraquinone compounds is of some industrial importance ((10.29): Y = OAlk; OAr).

Sulphonic groups can also be introduced oxidatively by treatment with sodium sulphite (cf. (10.29): Y = $SO_{33}^{\ominus}$). When sodium cyanide reacts with (10.56) (X = H), even with the exclusion of atmospheric oxygen, only the 2,3-dicyano derivative can be isolated; obviously the primary substitution product (10.62) (Y = CN) is oxidized by (10.56) and subsequently again reacts with cyanide; the second cyanide addition must occur more rapidly than the first, because of the electron-accepting effect of the cyano group already present.

Many of the reactions discussed above are reversible. Thus, halogen, sulphonic, nitro and hydroxy groups can be split off via the leuco compound, as halide, sulphite, nitrite and hydroxyl anions, respectively (cf. (10.26a)).

Leuco compounds of anthraquinones not only can split off nucleofugal groups, but also are capable of adding electrophilic groups. The oxidizing effect of oleum on hydroxyanthraquinones (cf. (10.18)) is attributable to this fact; this is also utilized in alkylation (10.30).

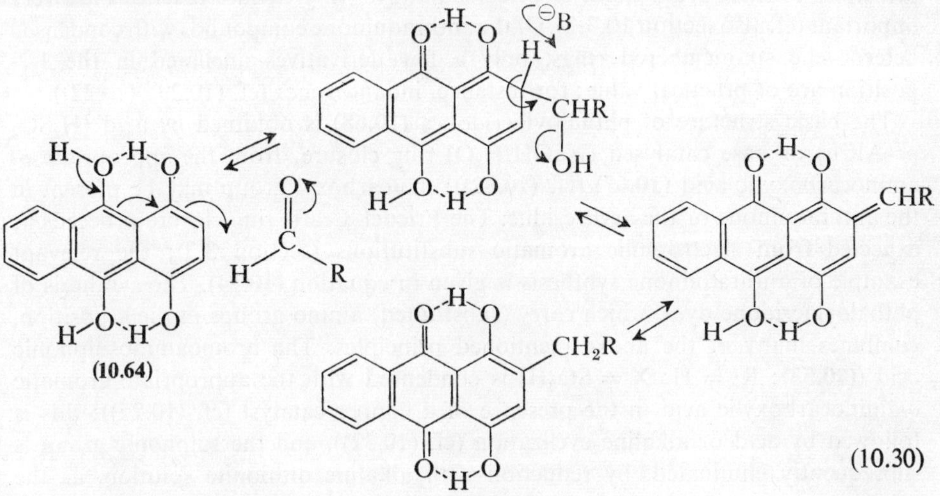

(10.30)

The aryl coupling in the preparation (10.31) of 2,2'-diamino-1,1'-dianthraquinone ((10.61): X = NH_2)† proceeds analogously to (10.23), (10.26) and (10.27). Here, owing to the formation of a complex with antimony pentachloride, the amino group loses its character as a nucleophilic reagent (cf. (10.21)).

In the preparation of anellated anthraquinone compounds, quite apart from the

† (10.61): X = NH^2 is a red pigment, whose crystals are constructed of parallel layers.[12]

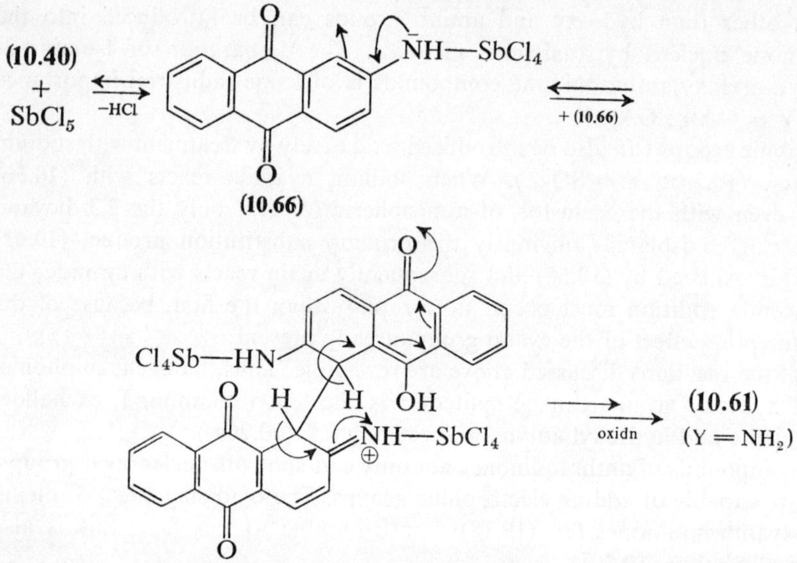

(10.40)
+ $\rightleftarrows$ [structure **(10.66)**] $\rightleftarrows$
$SbCl_5$ $-HCl$ $+ (10.66)$

[reaction scheme showing intermediate] $\xrightarrow{\text{oxidn}}$ **(10.61)** (Y = NH_2) (10.31)

principles of their production, as discussed above, ring closure reactions are very important (cf. also section 10.3.2). Of the anthraquinone compounds with condensed heterocyclic six-membered rings, only a few derivatives anellated in the 1,2-position are of practical value; for instance, indanthrones (cf. (10.20)–(10.22)).

The basic structure of phthaloylacridones **(10.68)** is obtained by acid (H_2SO_4 or $AlCl_3$) or base catalysed ($NaOH/H_2O$) ring closure, from the appropriate *o*-aminocarboxylic acid **(10.67)** (cf. (10.32)); the carboxyl group may be present in the anthraquinone or the aryl residue. The Friedel–Crafts ring closure proceeds as expected from electrophilic aromatic substitutions (section 3.3); the relevant example of anthraquinone synthesis is given in equation (10.34). The synthesis of phthaloylacridone dyes, which carry (substituted) amino groups in the 4-position, combines many of the above mentioned principles. The bromoaminosulphonic acid (**(10.53)**: R^1 = H; X = SO_3H) is condensed with the appropriate aromatic *o*-aminocarboxylic acid in the presence of a copper catalyst (cf. (10.25)); this is followed by acid or alkaline cyclization (cf. (10.32)), and the sulphonic group is subsequently eliminated, by reduction with alkaline dithionite solution, as the sulphite anion (analogously to the splitting off of hydrogen bromide with (10.26a)).

Ring closure to form phthaloylacridines (e.g. **(10.106)**) occurs according to the same principles (cf. (10.33)) used in the synthesis of phthaloylcarbazoles.

Diphthaloylcarbazole derivatives **(10.70)** are, industrially, the most important anthraquinone dyes with anellated heterocyclic five-membered rings. Oxidative ring closure to form the carbazole compound is normally the last step of the synthesis; it can be performed with sulphuric acid, with $AlCl_3$ in pyridine, or in an $AlCl_3$–NaCl melt with subsequent oxidation, e.g. with sodium bichromate (10.33).

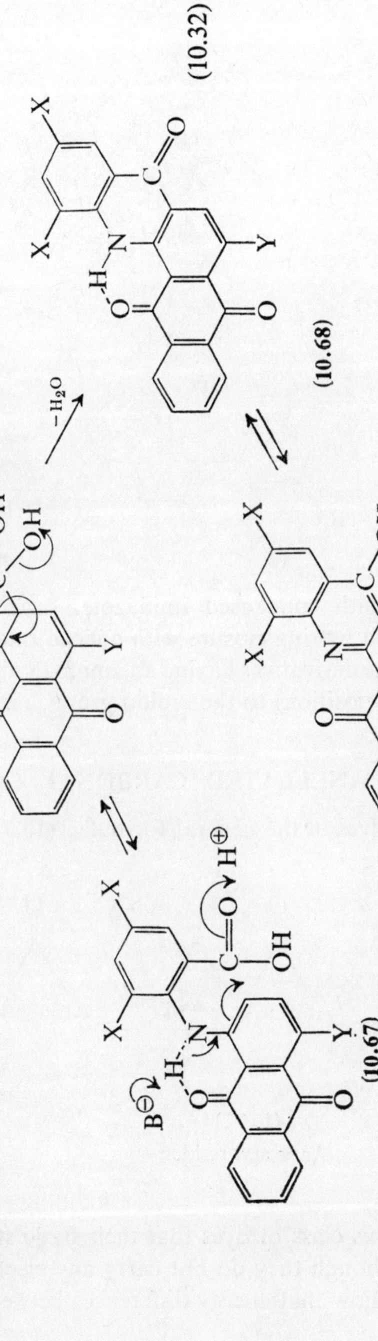

(10.32)

(10.68)

(10.67)

−H₂O

A radical mechanism has recently also been postulated for the cyclization.[11a] The most important representatives of this type of dye are prepared from 1,1'-anthrimides **(10.69)**.

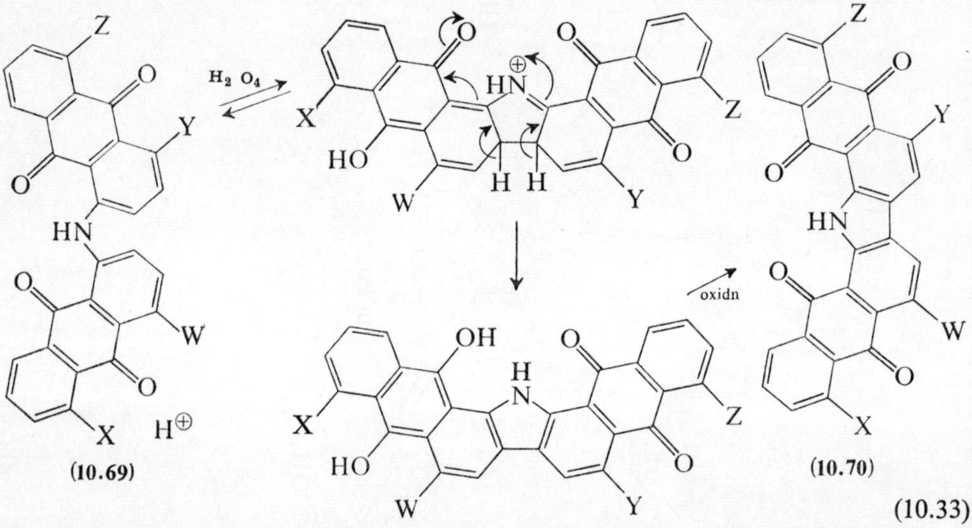

(10.69)   (10.33)   (10.70)

Anthraquinone dyes with condensed imidazole, oxazole and thiazole rings (cf. **(10.105)**) are produced by ring closure with concentrated sulphuric acid, from 2-arylamidoanthraquinone derivatives having an amino, hydroxy or mercapto group in the *o*-position (1- or 3-position) to the amido group, respectively.

10.3. HIGHER ANELLATED CARBONYL COMPOUNDS

In this section, carbonyl dyes of the general formulae **(10.71)**, **(10.72)** and **(10.73)** are discussed.

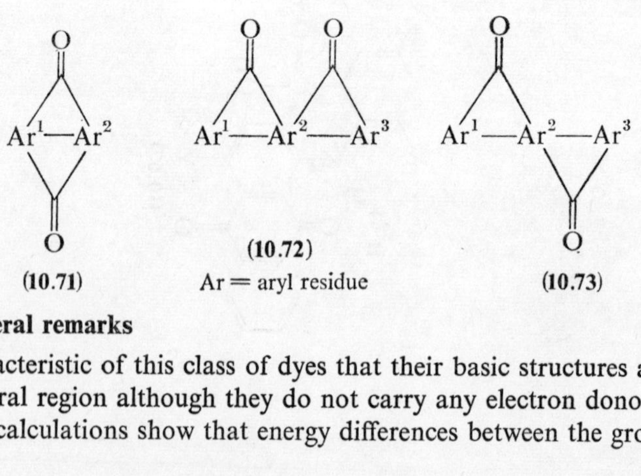

(10.71)   (10.72)   (10.73)

Ar = aryl residue

10.3.1. General remarks

It is characteristic of this class of dyes that their basic structures absorb in the visible spectral region although they do not carry any electron donors. Quantum mechanical calculations show that energy differences between the ground and the excited states (cf. section 2.1) are smaller as the size of the system of conjugated double bonds is increased (cf. also Chapter 6). Smaller excitation energies, however, denote absorption of light of a longer wavelength. The anellation of several aromatic rings can therefore shift the absorption spectrum from the ultraviolet to the visible region. Representatives of this class of dyes are used in textile dyeing solely as vat dyes (section 10.5.5).

10.3.2. Structures and principles of preparation

For the synthesis of polycyclic carbonyl compounds, the following types of ring closure reactions are available:
(a) Friedel–Crafts reactions of acylium ions,
(b) ring closures initiated by nucleophilic attack on activated aromatic C-atoms and subsequent cleavage of a substituent $R^{\ominus}$, or $R^{\oplus}$ + 2 electrons.

Friedel–Crafts ring closures proceed as expected for electrophilic aromatic substitutions (cf. section 3.3). An example is the preparation of anthanthrone ((**10.75**): X = H) from 1,1′-dinaphthyl-8,8′-dicarboxylic acid ((**10.74**): X = H) (cf. (10.34)). Instead of using $AlCl_3$, ring closure can be performed in sulphuric acid. Analogously, for instance, *cis*-(or *trans*-)dibenzpyrenequinone (**10.77**) or (**10.78**), pyranthrone ((**10.76**): X = CH), violanthrone (**10.86**) and isoviolanthrone (**10.87**) can be made from the appropriate carboxylic acids; however this method of synthesis is not used industrially.

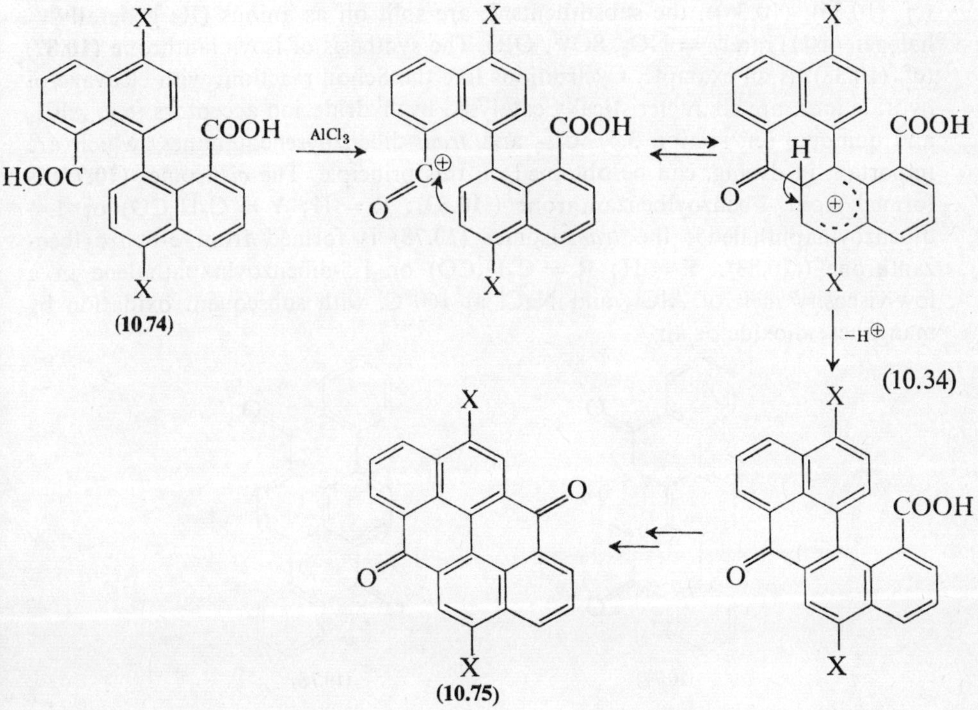

Ring closures initiated by nucleophilic attack are characteristic of the chemistry of carbonyl dyes. They include the syntheses (10.35) of pyranthrone ((**10.76**): X = CH) and flavanthrone ((**10.76**): X = N).

$$(10.61) \xrightarrow[-2H_2O]{OH^{\ominus}} (10.76) \qquad (10.35)$$

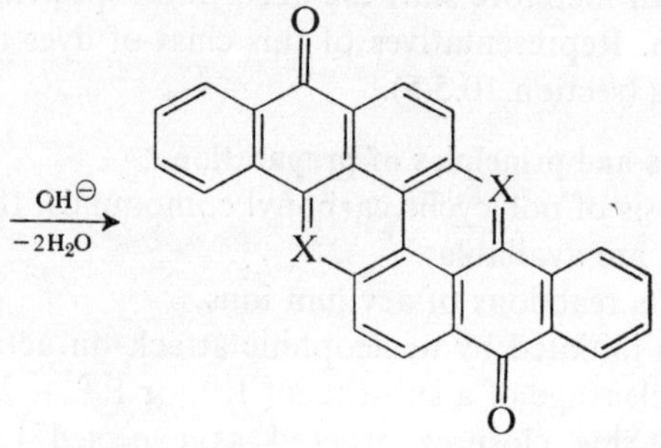

The corresponding dianthraquinone derivatives (**10.61**) can be prepared from 1-chloro-2-methylanthraquinones (for pyranthrone) or from 1-chloro-2-phthalimidoanthraquinone (for flavanthrone) in an Ullmann condensation (cf. (10.27)). In another technically important synthesis of flavanthrone, 2-aminoanthraquinone is treated with antimony pentachloride which, by complex formation, prevents anthrimide from forming (cf. (10.31)). With numerous ring closures of this kind (cf. (10.35), (10.38)), the substituents R are split off as anions (R: generally = halogen or H; rarely = NO_2, $SO_3^{\ominus}$, OR). The synthesis of isoviolanthrone (**10.87**) (cf. (10.38)) is an example. Cyclizations like the Scholl reaction, with cleavage of hydride ions, are characteristically catalysed by hydride ion acceptors (e.g. $AlCl_3$ and quinone) (cf. section 3.5). *Cis-* and *trans*-dibenzpyrenequinones, which are important in dyeing, can be obtained on this principle. The *cis*-isomer (**10.77**) is formed from 4-benzoylbenzanthrone ((**10.83**): R = H; Y = C_6H_5CO) or 1,4-dibenzoylnaphthalene; the *trans*-isomer (**10.78**) is formed from 3-benzoylbenzanthrone ((**10.83**): Y = H; R = C_6H_5CO) or 1,5-dibenzoylnaphthalene in a low-viscosity melt of $AlCl_3$ and NaCl at 160°C, with subsequent oxidation by manganese dioxide or air.

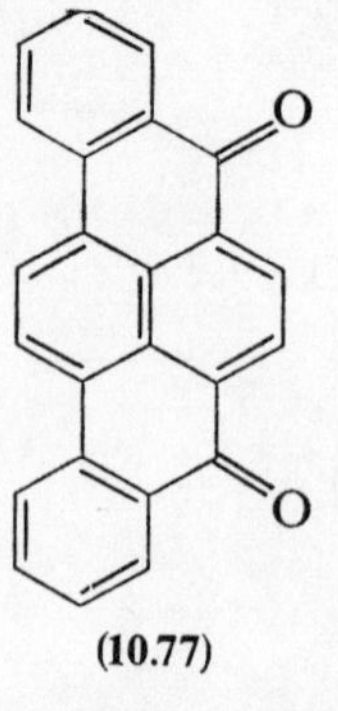

(10.77) 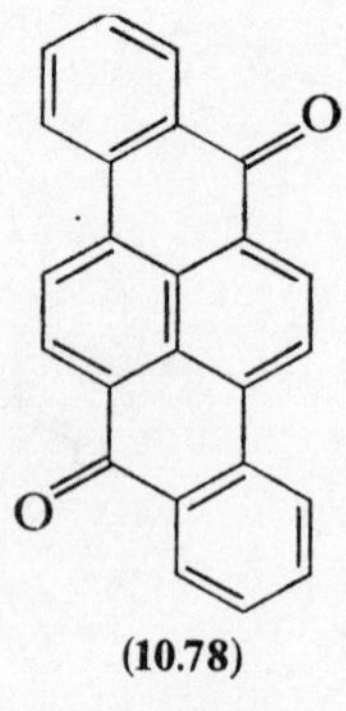   (10.78)

In cases where an intermediate addition product containing further acidifying substituents, e.g. carbonyl or nitro groups, is formed during ring closure, the hydrogen that is to be replaced can be split off as a proton, and the resulting intermediate product (usually in the leuco form) can be oxidized, by loss of two electrons, to the final product (cf. (10.37)). As regards mechanism, this reaction corresponds to the oxidative substitution of the aromatic hydrogen in anthraquinone chemistry (cf. (10.15), (10.20), (10.21), (10.22), (10.29) and (10.33)).

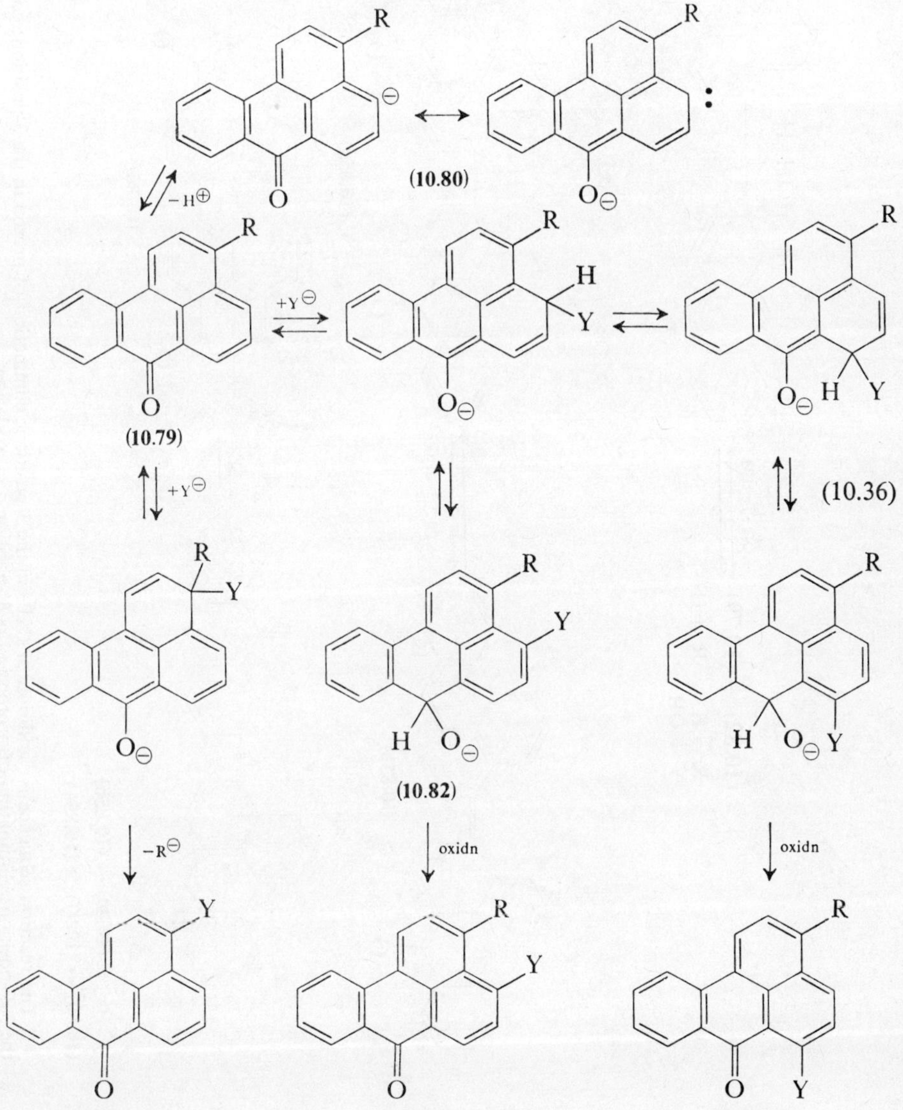

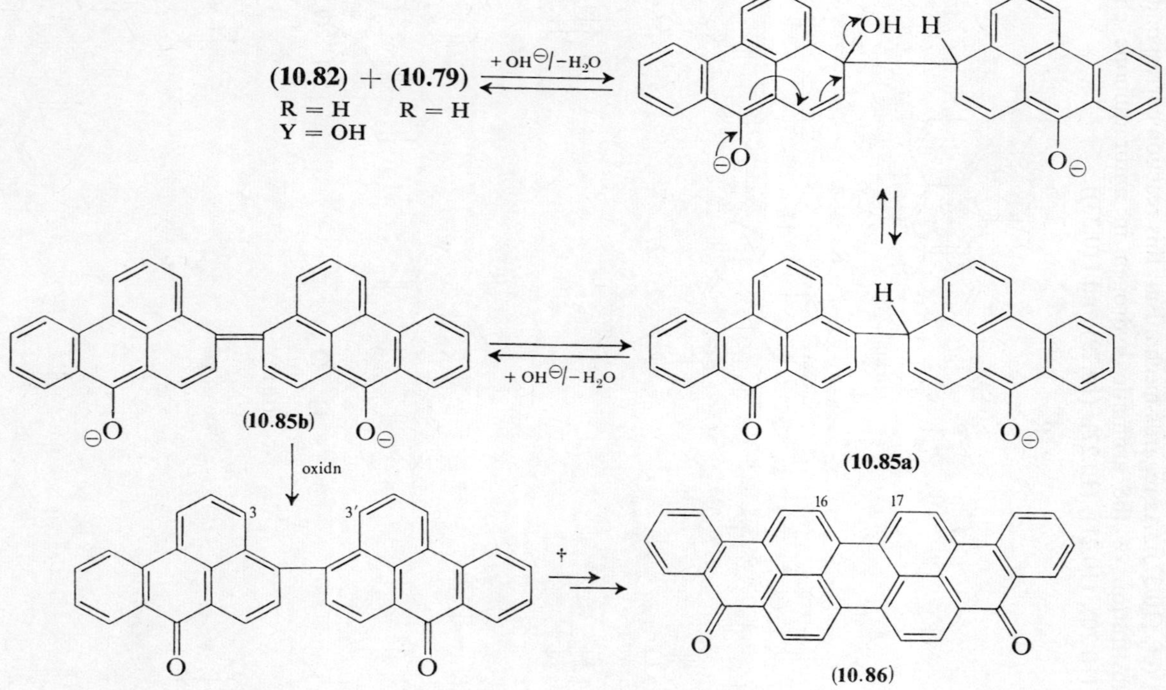

144   *Fundamentals of the Chemistry and Application of Dyes*

(10.37a)

(10.37b)
(10.37c)

(10.85a)

(10.86)

(10.85b)

(10.82) + (10.79)   R = H
R = H   Y = OH

2 (10.80) → (10.85b)
(10.80) + (10.97) ⇌ (10.85a)

† The reaction sequence is: addition of $OH^{\ominus}$ in the 3-position; elimination of proton in the 3-position; coupling of the aryl in the 3,3′-position; elimination of proton in the 3′-position and oxidation.

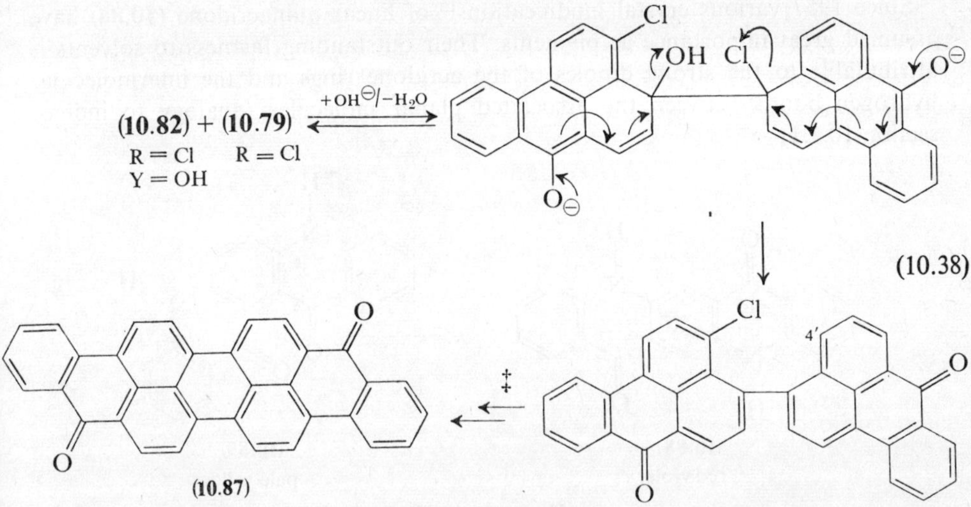

(10.82) + (10.79) $\xrightarrow{+OH^\ominus/-H_2O}$

R = Cl   R = Cl
Y = OH

(10.38)

(10.87)

The by-products occurring in the condensation reactions (10.37) and (10.38) are essentially derived from (10.36).

A KOH melt of benzanthrone ((10.79): R = H) produces, apart from violanthrone (10.86) as the main product, 4-hydroxybenzanthrone ((10.83): R = H; Y = OH) and traces of 6-hydroxybenzanthrone ((10.84): R = H; Y = OH). A similar result occurs when the reaction is performed with sodium anilide in aniline. However with sodamide in ammonia, 6-aminobenzanthrone ((10.84) R = H; Y = NH_2) is the main product.

So far, it has not been possible to decide whether violanthrone (10.86) is formed through dimerization of the carbene anion (10.80) (cf. (10.37b)), by reaction of (10.80) with (10.79) (cf. (10.37c)) or according to the mechanism given in (10.37a) (cf. reference 24).

From the point of view of dyeing, the halogen derivatives are the most valuable representatives of higher anellated carbonyl dyes. The halogen substituents are introduced by the customary methods of electrophilic substitution, mostly as the last step in the synthesis.

10.4. OTHER CARBONYL DYES

Apart from the dyes discussed in sections 10.1, 10.2 and 10.3, several smaller groups of conjugated dicarbonyl compounds exist, individual representatives of which are of industrial importance.

‡ The reaction sequence is: addition of $OH^\ominus$ in the 4'-position; elimination of proton in the 4'-position and nucleophilic substitution of $Cl^\ominus$.

Since 1957 various crystal modifications[12] of linear quinacridone (**10.88**) have assumed great importance as pigments. Their outstanding fastness to solvents is attributable to the strong dipoles of the acridone rings and the intermolecular hydrogen bonds between the associated planar molecules (analogy to indigo: section 10.1.1).

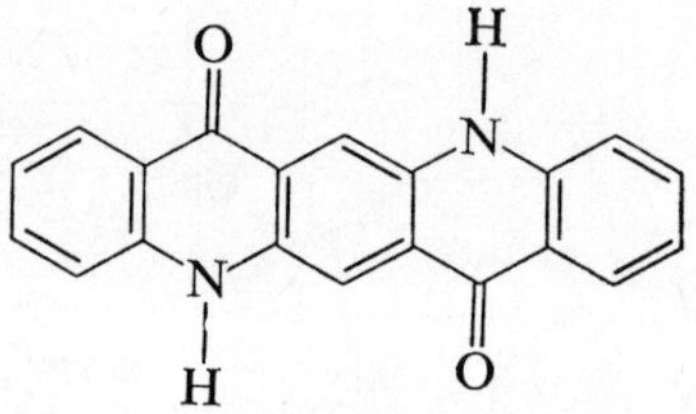

(**10.88**)
red violet

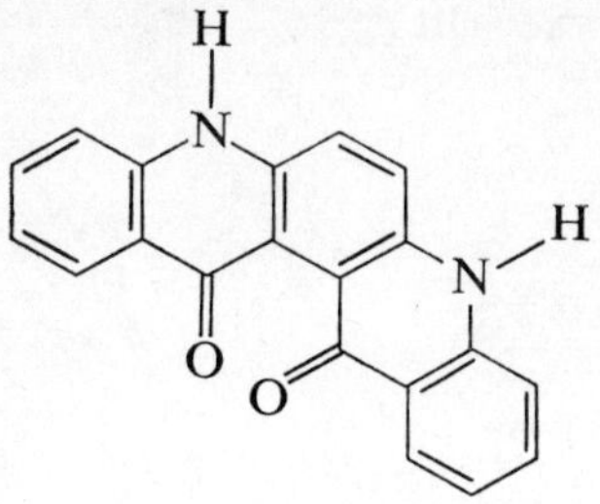

(**10.89**)
pale yellow

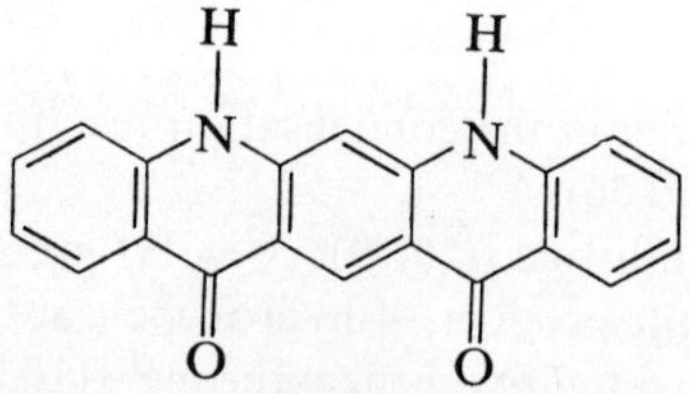

(**10.90**)
pale yellow

That (**10.88**) will absorb at longer wavelengths than (**10.89**) and (**10.90**) can be predicted from the rule of distribution of auxochromes (cf. section 2.1). A comprehensive account of the synthetic possibilities and the characteristics of quinacridones has been published.[13] Of the benzo- and naphthoquinone derivatives, only dyes derived from naphthazarin (**10.91**)[14] or diaminonaphthoquinone ((**10.93**): $R^1 = R^2 = X = H$) are industrially important. These are used as disperse dyes (cf. section 10.5.2).

Higher annellated carbonyl dyes, in which the vattable carbonyl groups are conjugated with other than aromatic double bonds (for aromaticity see the Hückel

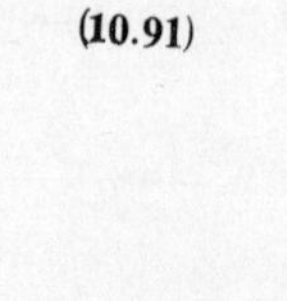

(**10.91**)

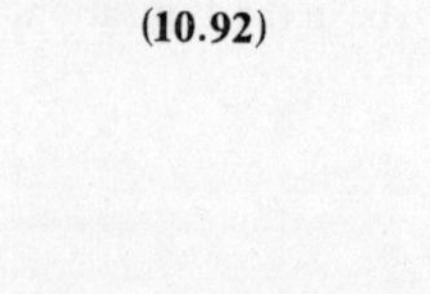

(**10.92**)

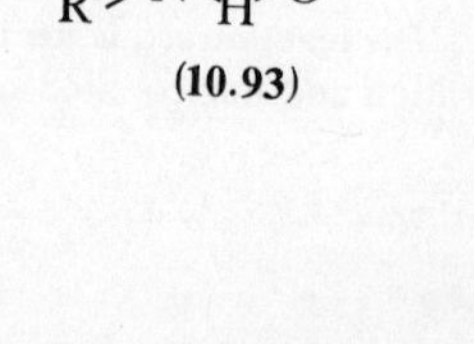

(**10.93**)

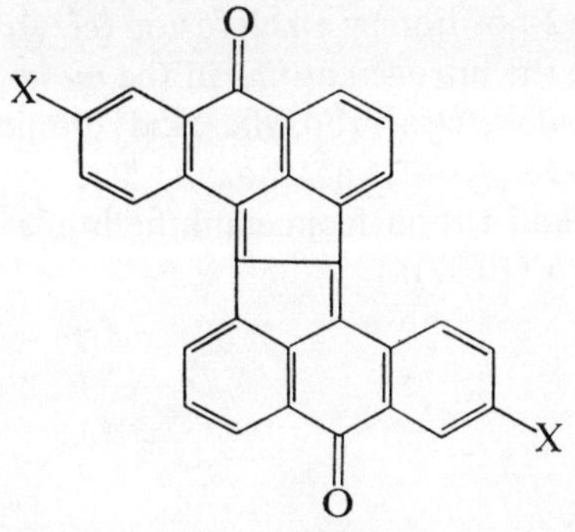

Carbonyl Dyes

(10.94)

rule, section 2.1), are derived from acedianthrone ((**10.94**): X = H). They are prepared by condensation of anthrone derivatives with glyoxal and subsequent ring closure with KOH or AlCl$_3$, by analogy with the mechanisms described in section 10.3.

Finally, some carbonyl compounds with heterocyclic rings containing nitrogen should be mentioned. These are anthrapyrimidines (**10.92**) that can be produced from the appropriate 1-aminoanthraquinone by condensation with formamide. The pyrazolanthrone derivatives (**10.96**), the diimides (**10.98**) and the diimidazoles (**10.100**) can be produced from aryl tetracarboxylic acids.

Pyrazolanthrone (**10.96**) could be defined as an inner azo compound (keto form); it can, for instance, be prepared by ring closure from 1-hydrazinoanthraquinone. It behaves similarly to benzanthrone (**10.79**: R = H) and, analogously (cf. (10.37)),

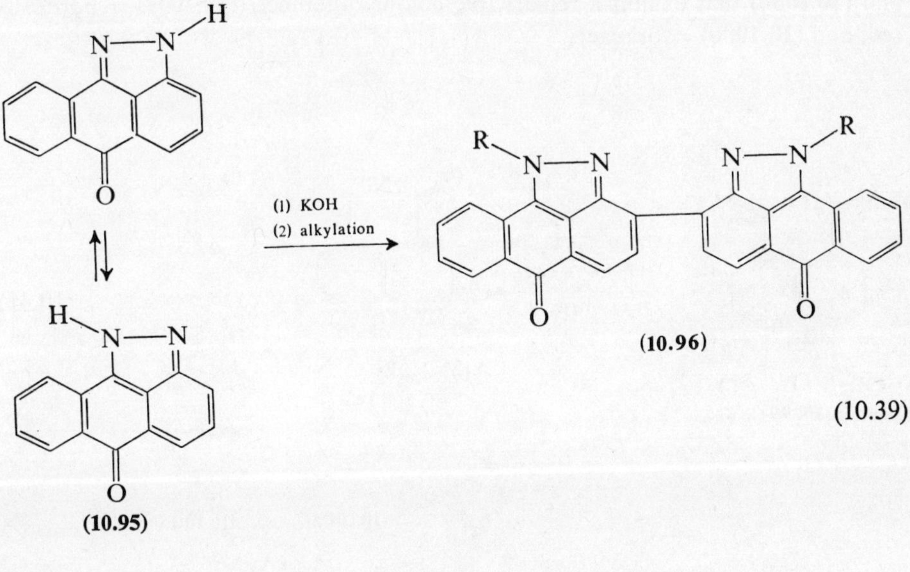

(1) KOH
(2) alkylation

(10.95)

(10.96)

(10.39)

it can be dimerized in the 2-position by alkali fusion (cf. (10.39)). In order to improve the resistance to alkalis, the nitrogen atoms in the *meso* position are alkylated.

The dyes of the perylenetetracarboxylic acid diimide series (**10.98**) can be prepared from perylene-3,4,9,10-tetracarboxylic acid by reaction with the appropriate amines, or by alkali fusion from naphthalimide (**10.97**) (10.40) (reaction mechanism, analogous to (10.37)).

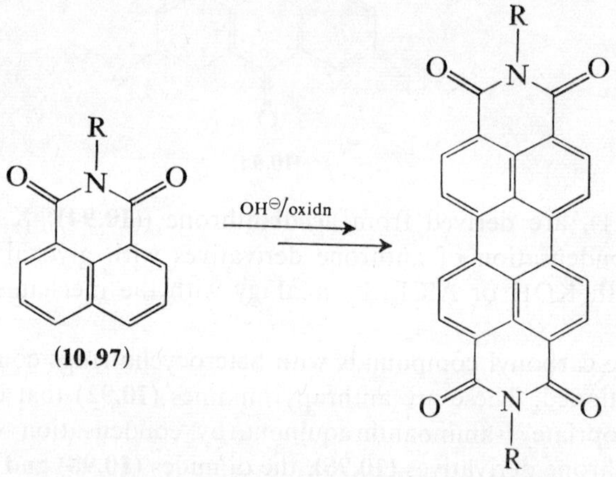

(10.40)

Analogous products are derived from naphthalene-1,4,5,8-tetracarboxylic acid anhydride (**10.99**). When R is an aromatic residue with an amino group at the *o*-position it can cyclize to give two isomeric diimidazole derivatives (**10.100a**) and (**10.100b**) that exhibit a remarkable colour difference ((**10.100a**) = bordeaux red, and (**10.100b**) = orange).

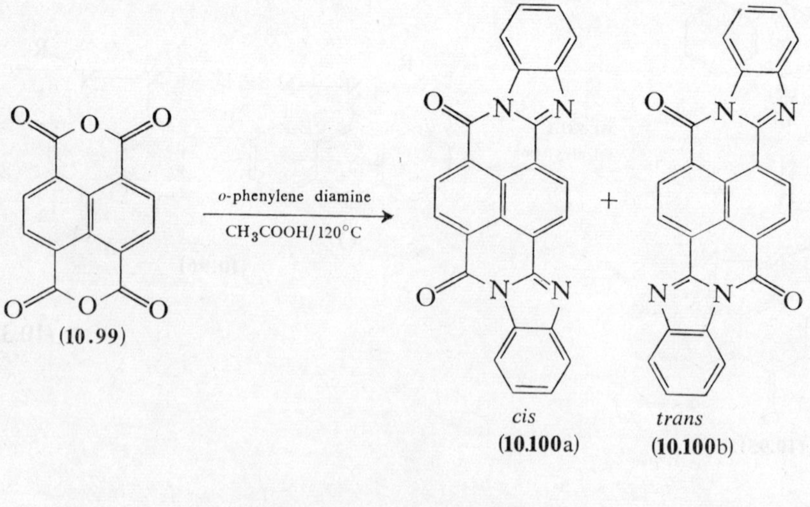

(10.41)

10.5. APPLICATION IN DYEING PROCESSES

10.5.1. Ionic carbonyl dyes

Ionic carbonyl dyes—mainly anthraquinone derivatives (section 10.2)—are characterized by the fact that they possess one or more water-solubilizing, ionizable substituents. With anionic dyes, these substituents are almost exclusively sulphonic acid groups, but with cationic dyes they are quaternary ammonium groups. The parent substances of anionic (acid) anthraquinone dyes are anthraquinone sulphonic acids with amino, substituted amino or hydroxyl groups in the α-positions. Their anionic character is of importance in dyeing protein fibres (cf. section 11.3.1). They are principally used for green, blue or violet hues. The sulphonic acid groups are either contained in the anthraquinone nucleus itself, an arylamino residue, or in the form of sulphuric esters at the aliphatic side chains. Products that are sulphonated at the nucleus, as, for instance, the wash-fast Carbolan Blue B ((**10.56**): $R^1 = H$; $R^2 = p$-(n-dodecyl-)phenyl; $X = SO_3H$; C.I. Acid Blue 138), are mainly derivatives of bromoamino sulphonic acid ((**10.53**): $X = SO_3H$). Substitution of the n-dodecyl residue gives rise to a characteristic change in the dyeing properties of this type of dye. With increasing length of the hydrocarbon chain, the water solubility drops and, in contrast, the affinity for protein fibres and hence fastness to washing rises owing to hydrophobic interactions with the protein (section 11.3.1). Analogously, with Alizarine Cyanine Green G Extra ((**10.52**): $R^1 = R^2 = p$-(2-sulpho-)tolyl; $X = Y = Z = H$; C.I. Acid Green 25), substitution of the methyl groups by butyl groups (to give Carbolan Green G) considerably increases fastness to washing.

Sulphuric ester derivatives of some disperse dyes containing β-hydroxy ethyl groups (cf. section 10.5.2) are frequently used for dyeing cellulose acetate (for example, (**10.52**): $X = Y = Z = H$; $R^1 = CH_2—CH_2—OSO_3H$; R^2 = phenyl).

The influence on dyeing behaviour of intramolecular hydrogen bonds between the carbonyl groups and the substituents in the α-positions has already been pointed out in section 10.2. In order to prevent colour changes in soda washing, hydroxyl groups in the β-position have to be alkylated or arylated, as, for example, in Carbolan Violet 2R ((**10.63**): $R^1 = H$; R^2 = 4-sulphophenyl; $Y = OC_{12}H_{25}$; C.I. Acid Violet 51). Here, too, the length of the alkyl group in the substituent Y may affect resistance to washing.

Chemically, the class of anionic carbonyl dyes includes those reactive dyes (section 10.5.4) that contain, apart from the usual structural features of this group of dyes, residues that can form covalent bonds with the substrate.

Cationic carbonyl dyes of low molecular weight can be used for dyeing acrylic fibres. Substituted amino- and hydroxyanthraquinone derivatives are mainly employed. The cationic group, generally quaternary ammonium, is linked to the coloured base structure by an aliphatic chain (e.g. Sandocryl dyes with trialkyl-hydrazinium groups).

10.5.2. Disperse dyes

25 per cent of all disperse dyes used at present are carbonyl compounds of the anthraquinone series with electron donors in the α-positions. The effect of the substituents on their absorption spectra (cf. Table 10.5) can be understood by taking into account the strength of the electron donors

$$(OH < NH_2 < NR_2 < NHAr)$$

and with the aid of the distribution rule of auxochromes (cf. section 2.1).

An optimum structure for a disperse dye should provide, apart from the greatest possible colour fastness (sublimation, light and wash fastness), an ideal relation between water solubility and hydrophobic properties during the dyeing process (section 11.3). However, since the structural requirements of many of these properties conflict, it is necessary to strike a compromise that will suit the application. Thus, for instance, adequate resistance to washing requires the highest possible standard affinity of the dye for the substrate, but this exerts an unfavourable effect on the water solubility necessary in dyeing; also dyeing speeds are closely connected with affinity (cf. section 11.3.2). The poor water solubility can be adjusted, as desired, e.g. by the introduction of β-hydroxyethyl groups. Celliton Fast Blue Green B ((**10.52**): Y = H; X = Z = OH; $R^1 = R^2 = CH_2$—CH_2—OH; C.I. Disperse Blue 7) is a typical representative. Polar substituents that improve water solubility generally raise fastness to sublimation. When these polar groups are also +M-substituents (electron acceptors, such as NO_2, CN or CONHR) this will also improve the resistance of these dyes to gas fumes. Irreversible colour changes caused by gas fumes are attributable to the diazotization of the amino groups by nitrogen oxides.

Light fastness depends on the structure of the dye, as well as the substrate. In order to discuss the influence of constitution on the fading mechanism, it is necessary to be in possession of exact data on the energy relations of excited states (cf. section 2.1). Since such data are lacking at the present time, one must revert to empirical rules.

For disperse dyes, in general, the rule that fastness to light is improved by reducing the basicity of the carbonyl group applies, provided that fading involves reduction of the dye. Basicity may be masked by the introduction of +M-substituents (halogen, COR, CN, NO_2), and by the formation of intramolecular hydrogen bonds.

Only a few representatives of other classes of carbonyl dyes have become technically important, as, for instance, Artisil Blue GLF ((**10.93**): $R^1 = R^2 = H$; X = Br; C.I. Disperse Blue 20) or Duranol Brilliant Yellow 6G ((**10.79**): R = OCH_3; C.I. Disperse Yellow 13).

The multiplicity of constitutional features of modern disperse dyes and the connection between constitution and colouristic properties have been discussed.[15,16]

Carbonyl Dyes

With reactive disperse dyes, epoxide and 3-chloro-2-hydroxypropyl groups are of importance. These react with the amino groups of polyamide fibres. Example: Procinyl Blue RS ((10.52): $X = Y = Z = H$; $R^1 = R^2 = CH_2-CHOH-CH_2Cl$; C.I. Reactive Blue 6).

10.5.3. Complex-forming carbonyl dyes

Carbonyl dyes that form complexes, the so-called mordant dyes, are of almost no practical importance, mainly because they are very troublesome to apply. Chemically, they are polyhydroxyanthraquinone derivatives with at least two hydroxyl groups in the *ortho* position. They are occasionally used as pigments and for dyeing polypropylene. Turkey Red, an aluminium calcium complex of alizarin (10.34), is of historical interest. A typical dye used for polypropylene fibres has the structure (10.52)($X = H$; $Y = Z = OH$; $R^1 = p$-(n-octyl-)phenyl; $R^2 = $ n-hexyl). The metal cations required for complex formation are nickel, cobalt, chromium and aluminium; these are applied either during or after application of the dye to the fibre.

10.5.4. Reactive dyes

In addition to a few reactive disperse dyes (cf. section 10.5.2) the great majority of reactive dyes of the carbonyl dye group are derivatives of bromoaminoanthraquinonesulphonic acid with the structure (10.101). The multiplicity of reactive groups is discussed in section 4.2.6.

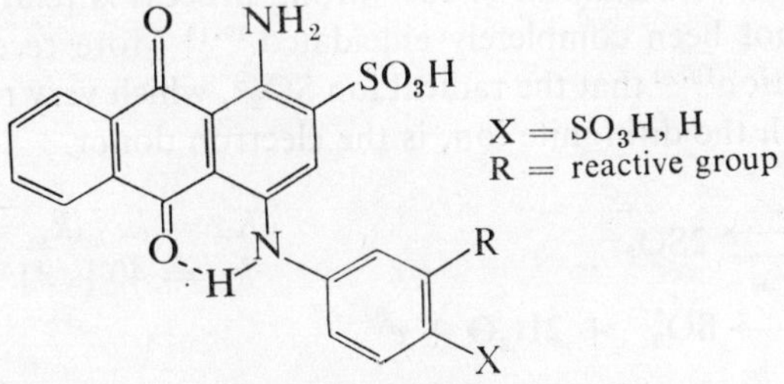

$X = SO_3H, H$
$R = $ reactive group

(10.101)

10.5.5. Vat dyes

Vat dyes are coloured carbonyl compounds that are practically insoluble in water but can be transformed by reduction (vatting) into a compound ('leuco form') soluble in aqueous alkali and dyed in this form. Air oxidation will re-form the original dye on the fibres. All vat dyes in use nowadays contain, as a characteristic grouping, a chain of conjugated double bonds with two keto groups in the end positions.†

† In exceptional cases (compare anthrapyrimidine (10.92)), a keto group may be replaced by another reducing group.

$$O=\overset{|}{C}-\!\!\left(\overset{|}{C}=\overset{|}{C}\right)_{\!n}\!\!-\overset{|}{C}=O \underset{\text{oxidn}}{\overset{\text{redn}}{\rightleftarrows}} HO-\overset{|}{C}=\!\!\left(\overset{|}{C}-\overset{|}{C}\right)_{\!n}\!\!=\overset{|}{C}-OH \quad (10.42)$$

dye   leucoform

$n = 1$: indigo, benzo-, naphtho- and anthraquinone
$n = 2$: anthanthrone, dibenzpyrenequinone, etc.
$n = 4$: pyranthrone, dibenzathrone, dipyrazolanthrone, etc.
$n = 5$: 4'.4''-dibenzanthronyl-1.2-ethylene derivative

Both carbonyl groups can be sited in the *trans*-position to the ethylene group (e.g. indigo, section 10.1) or—in cyclic compounds (e.g. anthraquinone derivatives, section 10.2)—in the *cis*-position to the polyene system. As diols ('vat acids') the leuco derivatives are very sparingly soluble in water, but, since the hydroxyl groups have an enol character, they are acidic ($pK_a = 9$–11) and dissociate in alkaline media to form soluble enolates.

Sodium dithionite (sodium hydrosulphite: $Na_2S_2O_4$) in alkaline solution generally serves as the reducing agent.

$$S_2O_4^{2\ominus} + 4OH^{\ominus} + O=\overset{|}{C}-\!\!\left(\overset{|}{C}=\overset{|}{C}\right)_{\!n}\!\!-\overset{|}{C}=O \longrightarrow \quad (10.43)$$

$$2SO_3^{2\ominus} + 2H_2O + {}^{\ominus}O-\overset{|}{C}=\!\!\left(\overset{|}{C}-\overset{|}{C}\right)_{\!n}\!\!=\overset{|}{C}-O^{\ominus}$$

The electrochemical mechanism of the vatting process is relatively complicated and, so far, has not been completely elucidated.[17,18] More recent investigations lead to the assumption[19,20] that the radical ion $SO_2^{\ominus}$, which very rapidly establishes an equilibrium with the dithionite ion, is the electron donor.

$$S_2O_4^{2\ominus} \underset{k_{-m}}{\overset{k_m}{\rightleftarrows}} 2SO_2^{\dot{\ominus}} \qquad \begin{array}{l} K_m = k_{-m}/k_m = 1.6 \cdot 10^9 \text{ [l/mol]} \\ k_m = 40 \text{ [s}^{-1}\text{]} \quad (25°C) \end{array}$$

$$SO_2^{\dot{\ominus}} + 2^{\ominus}OH \longrightarrow SO_3^{2\ominus} + 2H_2O + e \qquad (10.44)$$

For printing with vat dyes, it is customary to use addition compounds of sodium dithionite and formaldehyde (e.g. Rongalite C = $HOCH_2$—$SO_2^{\ominus}$ $Na^{\oplus}$) in place of sodium dithionite, because the reducing effect begins only when the temperature is raised; it can be catalysed at will by adding nitroarylsulphonic acids (e.g. *m*-nitrobenzenesulphonic acid). The mechanism of this catalysis is still unknown; it probably proceeds via the formation of azo and/or azoxy compounds that are capable of reducing the quinone system of the dye more rapidly.

Only a few of the many problems of vat dyeing will be mentioned here. With numerous vat dyes (e.g. indanthrone (**10.47**)), irreversible changes may occur when optimum reduction conditions are not strictly observed; the same applies to the oxidation step. Dyes containing acyl groups may partly hydrolyse when pH and vat temperatures are too high. Leuco derivatives may occasionally split off substituents, such as halogen, nitro, sulpho or hydroxy groups in the form of nucleofugal groups (cf. (10.26a)).

Three methods, the so-called hot, warm and cold dyeing methods are used for the application of vat dyes. The dyeing processes differ in temperature, hydroxyl ion and electrolyte concentration:

Hot dyeing method	50–60°C	5 g/l NaOH; no NaCl
Warm dyeing method	40–50°C	2 g/l NaOH; 10–20 g/l NaCl
Cold dyeing method	25–30°C	2 g/l NaOH; 15–25 g/l NaCl

The differences in the three processes are a consequence of the constitutional features of the dyes in question. The leuco anions of the hot-dyeing dyes are large particles with a planar structure, which, consequently, (cf. section 11.3.1) possess a relatively high standard affinity for cellulose fibres. Although increased temperatures shift all dyeing equilibria to the detriment of the fibres (cf. section 11.3.1), these dyes can nevertheless be applied at relatively high temperatures. No addition of salt is required. The high temperature raises the rate of diffusion, which is itself small, and prevents the formation of excessively large dye–dye associates in solution. The high hydroxyl ion concentration also indirectly counteracts dye association, since it shifts the hydroquinone acid-base equilibria (cf. section 3.5) in favour of the dibasic anions. Dyes suitable for the cold dyeing method are, in contrast, small molecules, the leuco forms of which exhibit a lower degree of association, diffuse more rapidly and have a small standard affinity, so that they have to be dyed at the lowest possible temperature, but at high ionic strength. The warm-dyeing dyes are intermediate.

As a rule, the vatting of indigo derivatives (section 10.1) produces a hypsochromic shift, whereas the vatting of anthraquinone (section 10.2) and higher anellated carbonyl derivatives (section 10.3) produces a bathochromic displacement, of their absorption spectra. In contrast to most vat dyes, many indigo dyes, because they can be vatted in weakly alkaline conditions, play a greater rôle in wool than in cotton dyeing. The affinity of leuco indigo for wool and cotton is only moderate, so that several applications of dye with intermediate oxidation are generally necessary.

Asymmetrical indigos, thioindigos and also halogenated indigo derivatives have a greater affinity.

The structurally most simple vat dyes are the acylaminoanthraquinones, used for dyeing yellow, orange, red and violet hues. Derivatives of 1,4- and 1,5-substituted acylaminoanthraquinones ((**10.102**)–(**10.104**)) are technically the most important. Because of their sensitivity to hydrolysis, dyes containing acyl groups are dyed by the cold process. That the azo bridge in (**10.102**) remains intact during the vatting process whilst, in general, azo dyes are reduced in these conditions, indicates that

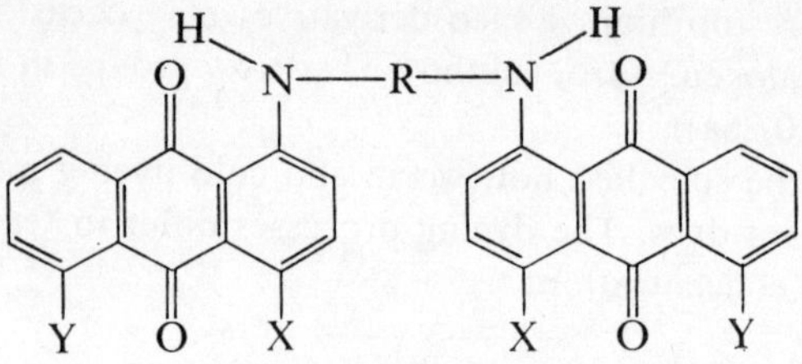

(**10.102**): Indanthrene Yellow 2GF: X = H; Y = NHCO—⟨phenyl⟩;

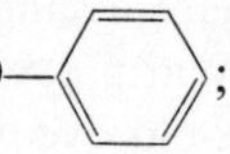

R = CO—⟨phenyl⟩—N=N—⟨phenyl⟩—CO

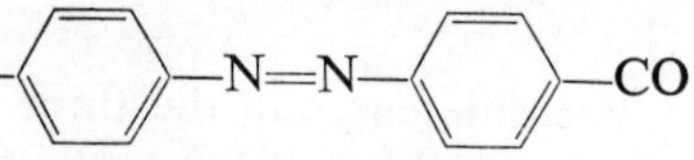

(**10.103**): Indanthrene Yellow 5GK: X = H; Y = H; R = ⟨m-phenylene⟩(OC—)(—CO)

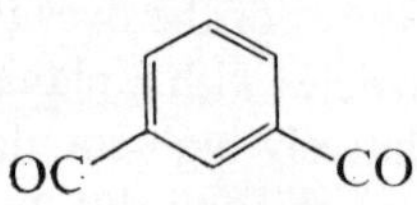

(**10.104**): Cibanone Red G: X = OCH_3; Y = H; R = ⟨4-amino-1,3,5-triazine-2,6-diyl⟩

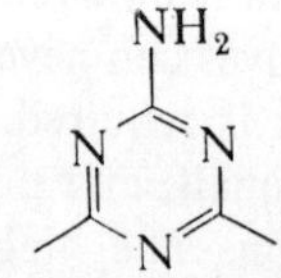

the formation of a tautomeric hydrazone form is necessary for the reduction of azo bridges;[21] this form of equilibrium is not present in (**10.102**).

The 1,1'-anthrimides (**10.69**) have lost their importance as dyes, but they are used as intermediate products in the preparation of dinaphthoylcarbazoles (**10.70**). Some of the most important anthraquinone vat dyes belong to this class. They are orange, olive, khaki and brown in colour, possess good overall properties and are almost exclusively applied by the hot dyeing process. Typical representatives are: Indanthrene Brown R ((**10.70**): W = Z = benzoylamino; X = Y = H; C.I. Vat Brown 3), and Indanthrene Olive R ((**10.70**): W = Y = benzoylamino; C.I. Vat Black 27). Some carbazole derivatives, made from anthrimides (synthesised from 1-aminoanthraquinone and the halogen derivatives of polycyclic aromatics) are also known. Dyes with several carbazole residues have dull hues. Indanthrene Khaki GG (C.I. Vat Green 8) is an example of great technical importance. It is made from 1,4,5,8-pentaanthrimide by carbazolization. Anthraquinone derivatives with condensed imidazole, oxazole and thiazole rings are of very limited importance as vat dyes. The oldest and most important representative of this class is Algol Yellow GC ((**10.105**); C.I. Vat Yellow 2), used for hot dyeing.

A large group of vat dyes are the phthaloylacridones (**10.68**). As a rule they are applied by the cold method, and have good affinity for cotton; their range of colours comprises orange, red, violet, blue, green and brown. Because of the keto/enol equilibrium (cf. (**10.68**)) they have only moderate resistance to boiling soda.

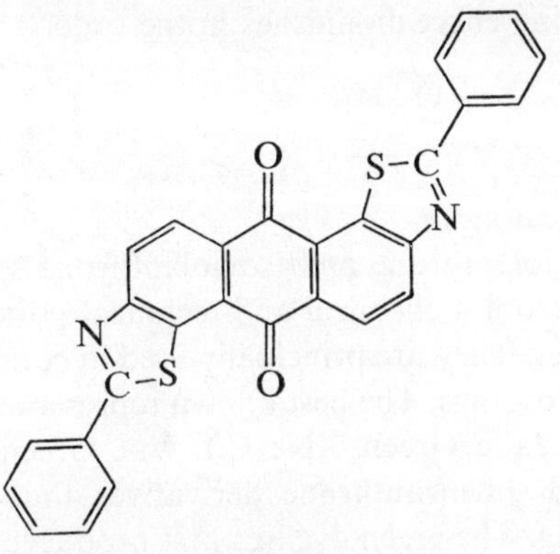

Carbonyl Dyes

(10.105)

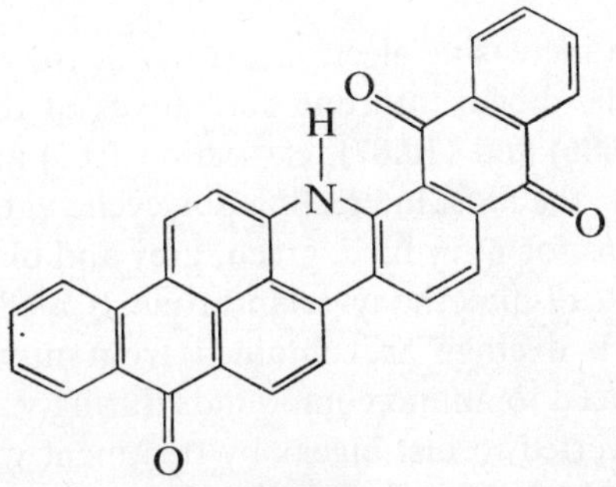

(10.106)

Two typical representatives are Indanthrene Red Violet RRK ((10.68): X = Y = Cl; C.I. Vat Violet 14) and Indanthrene Turquoise Blue 3GK ((10.68): X = Cl; Y = NH_2; C.I. Vat Blue 33). The thioxanthones and thioxanthenes, that are related to the acridones (cf. section 10.2.2), are nowadays unimportant. The phthaloylacridines (10.106) are a group of hot-dyeing vat dyes of excellent overall fastness. The parent substance is Indanthrene Olive Green B (10.106) (C.I. Vat Green 3).

Historically, as well as technically, indanthrones (10.47) are of great importance as vat dyes for cotton. When R. Bohn discovered indanthrone (10.47) in 1902, it was the first anthraquinone vat dye. The indanthrones are applied by the hot dyeing method; they possess good light fastness but are very sensitive to oxidizing (bleaching) agents since their dihydroazine rings are readily and reversibly oxidized to yellow azine derivatives. In a normal vatting process, only two carbonyl groups are reduced; more vigorous reduction will produce undesired side products. The parent substance, indanthrone ((10.47) Indanthrene Blue RS: C.I. Vat Blue 4), is principally used as a pigment. The different halogen, mainly chlorine, derivatives of indanthrone (10.47) are considerably more resistant to oxidation.

The halogen derivatives of anthanthrone ((10.75): X = Br: Indanthrene Brilliant Orange RK: C.I. Vat Orange 3), *trans*-dibenzopyrenequinone ((10.78): Indanthrene Yellow GK) and its halogen derivatives are cold dyeing dyes used mainly for printing.

Flavanthrone ((10.76): X = N) and the isosteric pyranthrone ((10.76): X = CH) are hot dyeing dyes having good overall fastness. The substitution of two methine groups of pyranthrone by nitrogen in flavanthrone ((10.76): X = N; C.I. Vat Yellow 3) considerably reduces its fibre damaging effect. Fibre damage generally occurs in the redox reactions between dye and substrate. In this case the photochemically-excited (in the singlet or triplet state) dye molecules oxidize the cellulose fibres by abstracting hydrogen. The oxidizing effect diminishes in the order

$$\overset{\cdot}{\underset{/\,\backslash}{C}} > \overset{\cdot}{\underset{/\,\backslash}{N}} > -\overline{\underline{O}}\cdot$$

with increasing electron density at the radical atom.

The dibenzanthrone derivatives of the violanthrone and isoviolanthrone series ((**10.86**) and (**10.87**), cf. section 10.3) are, from a chemical and technical point of view, the most interesting polycyclic vat dyes. They are principally used in combinations for navy blue, green, grey and black dyeings. The best known representative is 16,17-dimethoxyviolanthrone (Caledon Jade Green XN; C.I. Vat Green 1). Black dyeings are obtained from nitrated dibenzanthrone derivatives that are reduced to amino compounds during vatting. The green dyeings first produced are converted to fast blacks by treatment with hypochlorite.

Only a few of the representatives of the carbonyl dyes mentioned in section 10.4 are used as vat dyes; Indanthrene Yellow 4GK ((**10.92**): $X^1 = H$, $X^2 = 2,4$-dichlorobenzoylamino; C.I. Vat Yellow 31), Indanthrene Rubine R ((**10.96**): $R = C_2H_5$; C.I. Vat Red 13), Indanthrene Red Brown 3R ((**10.94**): $X = Cl$) as representatives of the anthrapyrimidine, pyrazolanthrone and acedianthrone series, respectively, and the dyes Indanthrene Bordeaux RR ((**10.100a**); C.I. Vat Red 15), Indanthrene Brilliant Orange GR ((**10.100b**); C.I. Vat Orange 7), Indanthrene Red GG ((**10.98**): $R = CH_3$; C.I. Vat Red 23), which are derivatives of aryl tetracarboxylic acids, are all distinguished by their good general fastness.

10.5.6. Leuco sulphuric ester dyes

The application of vat dyes (section 10.5.5) has the disadvantage that they first have to be reduced to a water-soluble form. Indigosols are sulphuric esters of the leuco form of vat dyes and as such they are water-soluble. These sulphuric esters have affinity for cellulose; they are padded on to the fibre and oxidized to their parent dyes. Indigoid, as well as anthraquinoid vat dyes, can be converted to the leuco esters, e.g. Indigosol O ((**10.107**); C.I. Solubilized Vat Blue 1).

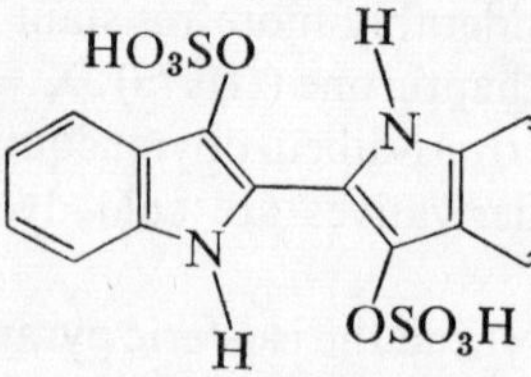

(**10.107**)

Reduction and esterification are generally carried out, without isolating the leuco compound, in pyridine with iron as reducing agent and chlorosulphonic acid as esterifying agent (10.45)–(10.47).

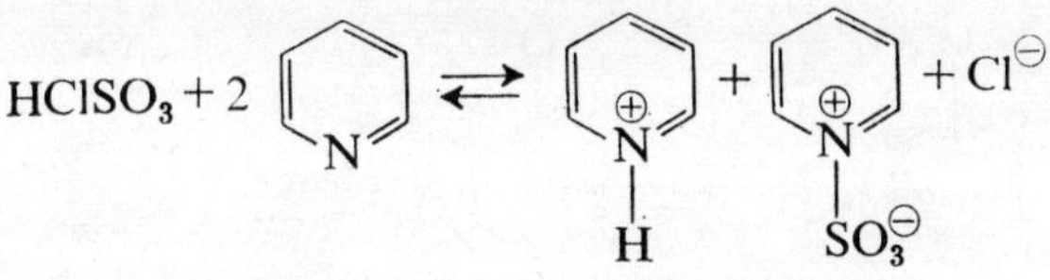

Carbonyl Dyes

$$HClSO_3 + 2 \; \text{Py} \rightleftharpoons \text{PyH}^+ \text{(10.108)} + \text{PySO}_3^- \text{(10.109)} + Cl^- \quad (10.45)$$

$$2(\mathbf{10.108}) + Me^{m\oplus} + O{=}C{-}(C{=}C)_n{-}C{=}O \longrightarrow$$
$$HO{-}C{=}(C{-}C)_n{=}C{-}OH + 2\,\text{Py} + 2Me^{(m+2)\oplus} \quad (10.46)$$ 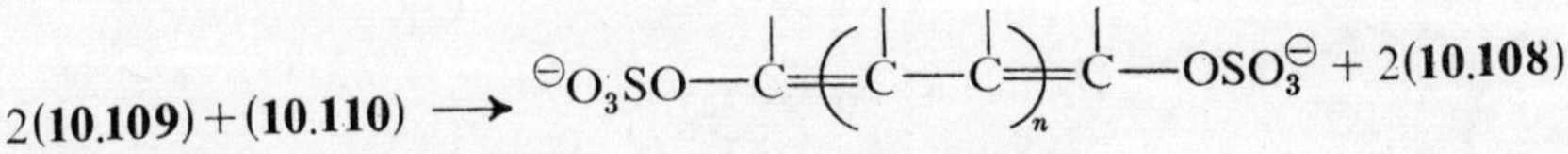
$$\mathbf{(10.110)}$$

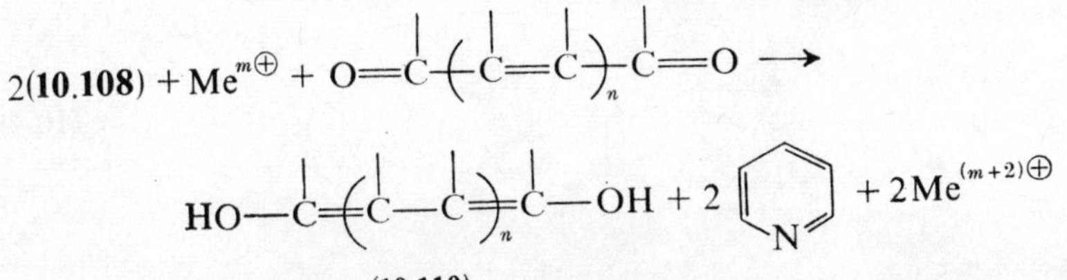

$$2(\mathbf{10.109}) + (\mathbf{10.110}) \longrightarrow {}^{\ominus}O_3SO{-}C{=}(C{-}C)_n{=}C{-}OSO_3^{\ominus} + 2(\mathbf{10.108}) \quad (10.47)$$ 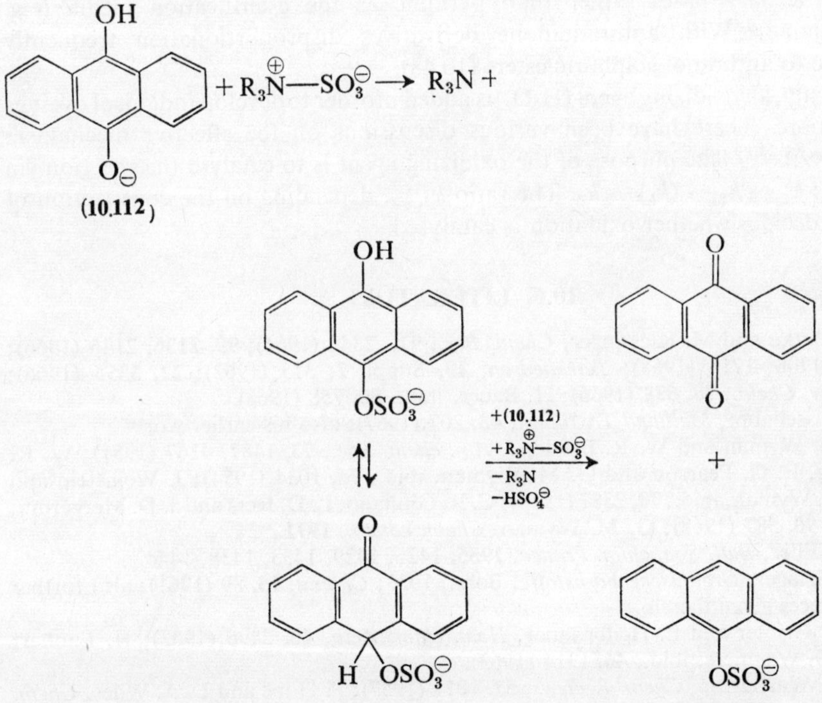
$$\mathbf{(10.111)}$$

(10.112): 9-hydroxy-10-oxido-anthracene + $R_3\overset{\oplus}{N}{-}SO_3^{\ominus} \longrightarrow R_3N\,+$

$$\begin{array}{c}\text{9-OH, 10-OSO}_3^{\ominus}\text{-anthracene} \\ \updownarrow \end{array} \xrightarrow[\substack{+R_3\overset{\oplus}{N}{-}SO_3^{\ominus} \\ -R_3N \\ -HSO_4^{\ominus}}]{+(\mathbf{10.112})} \text{anthraquinone} + \text{(10.113)} \quad (10.48)$$

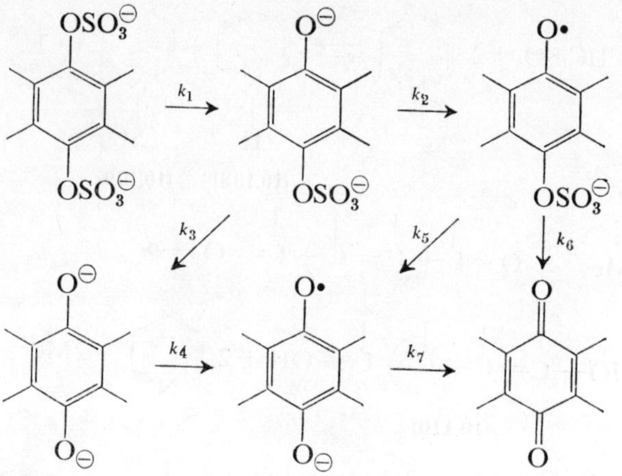

Hydrolysis: k_1; k_3; k_5
Oxidation: k_2; k_4; k_6; k_7

In some cases, Indigosols can be produced in aqueous media from the leuco forms of vat dyes; however, here it is necessary to use the N-sulphonic acids of stronger tertiary bases rather than pyridine as the esterification media (e.g. triethylamine). With anthraquinone derivatives disproportionation frequently gives rise to anthranol sulphuric esters (10.48).

Normally, an oxidizing agent (H_2O_2) is added in order to develop indigosol dyeings on the fibre. There have been various discussions on the effective mechanism (cf. (10.49)).[22,23] The purpose of the oxidizing agent is to catalyse the reaction via the path: $k_1 \rightarrow k_2 \rightarrow (k_5) \rightarrow k_6$. The ratio k_2/k_3, depending on the constitution of the dye, decides whether oxidation is catalysed.

10.6. LITERATURE

1. W. Lüttke and M. Klessinger, *Chem. Ber.*, **97**, 2342 (1964); **99**, 2136, 2146 (1966); **101**, 1708, 1715 (1968); *Tetrahedron*, **19**, Suppl. 2, 315 (1963); **22**, 3355 (1966); *Angew. Chem.*, **78**, 638 (1966); H. Bauer, ibid., **80**, 758 (1968).
2. H. Cassebaum, *Melliand Textilber.*, **48**, 207 (1967); also his earlier work.
3. G. M. Wyman and W. R. Brode, *J. Am. chem. Soc.*, **73**, 1487, 4267 (1951); W. R. Brode, E. G. Pearson and G. M. Wyman, ibid., **76**, 1034 (1954); J. Weinstein and G. M. Wyman, ibid., **78**, 2387 (1956); C. R. Giuliano, L. D. Hess and J. D. Mergerum, ibid., **90**, 587 (1968); G. M. Wyman, *Chem. comm.*, **1971**.
4. H. v. Eller, *Bull. Soc. chim. France*, **1955**, 1426, 1429, 1433, 1438, 1444.
5. R. Wizinger, *Organische Farbstoffe*, Bonn, 1933; *Chimia*, **15**, 89 (1961); also further references given therein.
6. cf. R. Grinter and E. Heilbronner, *Helv. chim. Acta*, **45**, 2496 (1962); H. Labhart and G. Wagnière, ibid., **46**, 1314 (1963).
7. W. K. Warburton, *Chem. Reviews*, **57**, 1011 (1957); P. Hope and L. A. Wiles, *Chem. and Ind.*, **1966**, 32.

8. G. A. Olah, *Friedel-Crafts and Related Reactions*, Vol. 4, Interscience, New York, 1963–1965.
9. H. Labhart, *Chimia*, **15**, 20 (1961); cf. also *Helv. chim. Acta*, **40**, 1410 (1957).
10. A. Kuboyama, *Bull. chem. Soc. Japan*, **31**, 752 (1958); see also ibid., **39**, 1874 (1966).
11. D. Hall and C. L. Nobbs, *Acta crystallogr.* [Copenhagen], **21**, 927 (1966); M. Bailey and C. J. Brown, ibid., **22**, 392, 488, 493 (1967).
11a. S. V. Sunthankar and R. Gopalan, *J. Soc. Dyers Colourists*, **85**, 373 (1969).
12. K. Ogawa, H. J. Scheel and F. Laves, *Naturwissenschaften* **53**, 700 (1966).
13. S. S. and L. L. Labana, *Chem. Reviews*, **67**, 1 (1967).
14. E. Merian, *Chimia*, **13**, 181 (1959).
15. C. Müller, *Supplementum Chimia*, **1968**, 69.
16. Y. Banshi, *Kogyo Kagaku Zassi*, **55**, 666 (1952).
17. B. Milicevic and G. Eigenmann, *Helv. chim. Acta*, **46**, 192 (1963).
18. S. Lynn, R. G. Rinker and W. H. Corcoran, *J. physic. Chem.*, **68**, 2363 (1964).
19. B. Milicevic, *Textilveredlung*, **1**, 84 (1966).
20. U. Baumgarte, *Textilveredlung*, **2**, 896 (1967); L. Burlamacchi, G. Guarini and E. Tiezzi, *Trans. Faraday Soc.*, **65**, 496 (1969).
21. M. Schellenberg and R. Steinmetz, *Helv. chim. Acta*, **52**, 431 (1969).
22. H. Schenkel, *Chimia*, **15**, 203 (1961).
23. A. Johnson and M. L. Rahman, *J. Soc. Dyers Colourists*, **74**, 291 (1958); A. Johnson and A. P. Lockett, ibid., **76**, 412 (1960); S. Ainsworth and A. Johnson, ibid., **71**, 592 (1955); **73**, 41 (1957).
24. H. W. Wanzlick and H. Ahrens, *Liebigs Ann. Chem.*, **693**, 176 (1966); W. Bradley and F. K. Sutcliffe, *J. chem. Soc.* (London), **1954**, 708; W. Bradley, *J. Soc. Dyers Colourists*, **70**, 57 (1954).
25. T. Furuya, S. Shibata and H. Lizuka, *J. Chromatogr.* [Amsterdam], **21**, 116 (1966).
26. J. Winkler and W. Jenny, *Helv. chim. Acta*, **48**, 119, 190 (1965).
27. W. Bradley and E. Leete, *J. chem. Soc.* (London), **1951**, 2129, 2147; W. Bradley, E. Leete and D. S. Stephens, ibid., **1951**, 2158, 2163; W. Bradley and H. E. Nursten, ibid., **1951**, 2170, 2177; **1952**, 3027; **1953**, 924; W. Bradley, R. F. Maisey and C. R. Thitchener, ibid., **1954**, 272.

11

The Application of Dyes

11.1. GENERAL REMARKS

THE dyes, and methods of application, must be chosen to suit the substrate and to give the required fastness properties. The development of new substrates presents new problems, gives rise to new dyes and dyeing methods and thus influences dyeing technology in a decisive manner. From the point of view of application two basic processes will be discussed—pigment dyeing (cf. section 11.2), which is based on a *mechanical anchoring* of the dye in the substrate, and exhaustion dyeing, which is based on *diffusion and adsorption processes* (cf. section 11.3), and may often be accompanied by chemical reactions (application of vat, reactive, Naphtol AS and chrome dyes).

Dyeing methods as a whole are extremely complicated processes, the deeper understanding of which requires, for example, a knowledge of the physical chemistry of diffusion and adsorption processes, of the physical and chemical characteristics of the dyes and of the substrates to be dyed. Despite the complexity of the dyeing systems, the discussion on dyeing mechanisms in this Chapter will be kept on an elementary, but still meaningful, level.

11.2. DYEING WITH ORGANIC PIGMENTS

Organic pigments are dyes that are insoluble in water and dissolve with great difficulty in organic solvents. They can be applied as metal-free, neutral compounds, as metal chelate complexes or as sparingly-soluble salts (lakes).

Although some of the most valuable *metal-free pigments* are found in the group of carbonyl dyes (Chapter 10) most of them (about 85 per cent) are azo compounds. Azo pigments are mainly yellow, orange (coupling products of acetoacetarylides or pyrazolones) and red (condensation products from coupled 2-naphthol-3-carboxylic acid derivatives and aromatic amines). The representatives of the carbonyl dye series (cf. Chapter 10) that must be mentioned are mainly the thioindigo derivatives (e.g. 5,5′, 8, 8′-tetrabromothioindigo), the acylaminoanthraquinones, the quinacridones (cf. **(10.88)**), the derivatives of the tetracarboxylic acids of naphthalene and perylene (cf. **(10.100)**, and **(10.98)**) as well as some derivatives of anthanthrone (**(10.75)**: $X = H$), flavanthrone (**(10.76)**: $X = N$), violanthrone (**(10.86)** and **(10.87)**) and indanthrone **(10.47)**. The dioxazines form a small but not insignificant group of metal-free violet pigments.

The insoluble phthalocyanines (cf. Chapter 7), which supply the fastest blue and green pigments, are *metal chelate complexes*.

Precipitation of water-soluble dyes as *sparingly soluble salts* produces lakes. The main precipitants for anionic dyes are alkaline earth and other bi- and higher valent metal salts; for cationic dyes heteropolyacids are used (Fanal dyes: cf. also section 8.3). Lakes are produced mainly from azo dyes but anthraquinone, triphenylmethane and thiazine dyes are also used.

The physical properties—crystal form, particle size, dispersibility, etc.—of organic pigments exert a decisive influence on their technical application. Since crystal form and particle size affect the hue and intensity of dye, their stability decides the possible uses of a pigment. The stable and metastable crystal forms of phthalocyanines have been studied comprehensively with reference to problems of application (cf. section 7.3). The hardness and size of pigment particles which are determined by the type of agglomerate formed during preparation and drying influence the abrasion resistance of prints. It is very difficult to grind and disperse particles that are too hard. Dispersibility influences the stability of suspensions of pigments in the appropriate media of application; it can be improved by the addition of dispersing agents. The dye chemist endeavours to synthesize pigments in a suitable crystalline form and having the optimum particle size (0·05–0·5 μm according to end-use) whenever possible. He is able to influence these features by careful selection of the reaction conditions (temperature, reaction medium), drying conditions, addition of dispersing agent, etc. Another, interesting, possible means of changing the physical properties of pigments is in the choice of the reaction sequence (e.g. condensation before azo coupling instead of azo coupling before condensation).

Organic pigments are commercially available as dispersions (pastes) or easily dispersible powders, tablets or flakes. These preparations always contain additives appropriate to the end-use. Thus, pigments for spin dyeing (spinning pigments) should be prepared in such a manner that they disperse as completely as possible in the spinning mass, leaving no aggregates; additives should not have any detrimental effect on the spinning process.

Disperse pigments for textile printing and dyeing contain a binder to anchor the pigment to the substrate. In practice, nearly all synthetic resins which are soluble in water or organic solvents, or can be dispersed in water, and which, in addition, have the capacity to form a film which adheres sufficiently strongly to, and covers, the pigment and fibres, can be used as binders. In modern pigment printing, emulsion printing pastes are mainly used; these may be either of the water-in-oil or of the oil-in-water type: in the first case, the pigments are dispersed in water, and in the second case, they are dispersed in organic solvents. Thickeners and binders may be dissolved or dispersed in either the aqueous or the organic phase.

Several works give a comprehensive discussion on pigments: general reviews,[1] chemistry and prospects,[2] physical chemistry,[3] applications.[1,4]

11.3. DYEING METHODS BASED ON EQUILIBRIUM PROCESSES

With the exception of pigment dyeing (cf. section 11.2) all dyeing methods are dependent on sorption processes, preceded by transport phenomena (mainly diffusion) and frequently accompanied by chemical reactions. A more exact description of these dyeing systems is extremely difficult and, if based solely on the three fundamental theories of physical chemistry, i.e. quantum mechanics, statistical mechanics and thermodynamics, tends to be rather abstract. Models† are required for the interpretation of the experimentally-determined phenomena. The more detailed the construction of these models, the better they agree with the results of experiments, but the more complicated are the mathematical expressions, so that eventually approximation processes (i.e. new hypotheses) are required for their solution, and this again simplifies the model. This means, of necessity, that there is a limit (Pauling point) to any arbitrary refinement and generalization of models to improve their agreement with experiment. It is, therefore, not surprising that so far no dyeing model ('dyeing theory') of general validity has been developed, i.e. which gives an unequivocal description of all dyeing phenomena.[4a]

Despite these shortcomings some of the most important characteristics of dyeing systems will be discussed on the basis of simple, but still meaningful, models. Such a procedure is, so far, the only way of giving dye chemists engaged in experimental work a useful (at best) semi-quantitative description of dyeing processes which can be used for the logical planning of experiments. Here it must be pointed out emphatically that, unfortunately, trade literature frequently uses ill-defined terms; this often makes a theoretical evaluation of the experimental dyeing results difficult or even impossible (cf. reference 5).‡

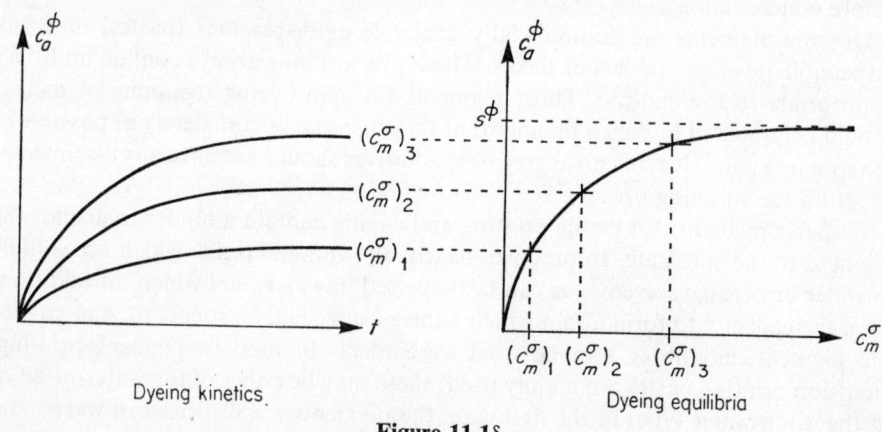

Dyeing kinetics Dyeing equilibria

Figure 11.1§

† Such models are in reality only hypotheses and can be called theories only in the widest meaning of the word (e.g. 'dyeing theories').
‡ In the following, some of these terms, e.g. affinity, reversibility, diffusion coefficient, etc. will be defined in the appropriate places.
§ For a definition of the terms cf. page 164.

The dyeing process, i.e. the distribution of a dye between at least two phases (dyeing bath and substrate) can be described (see Figure 11.1) by dyeing kinetics (transport and reaction phenomena: irreversible processes,‡ cf. section 11.3.2) and dyeing statics (sorption and desorption processes in the state of equilibrium: reversible processes,‡ cf. section 11.3.1).

11.3.1. The dyeing system in the state of equilibrium

The relation between the concentrations of a substance that is distributed, at constant temperature, between two phases is termed the *distribution isotherm*. Fundamentally, it does not matter whether one of the concentrations can be represented explicitly and mathematically as a function of the other or only graphically (Figure 11.1). It is customary to assume tacitly that these are always equilibrium concentrations. The distribution process is called *adsorption* if the substance which is to be distributed is retained by a surface (e.g. gas on a solid) but *absorption* if this occurs in the interior of a body (e.g. gas in liquid). When it is not known where the particles are retained initially and despite this fact it is to be emphasized that they originate from the other phase, the process is termed *sorption*. Thus the expression 'sorption isotherm' means simply 'distribution isotherm' and here it is not stated whether the *sorbate* (e.g. dye) entering from the surroundings (e.g. dyeing bath) is present on the surface or the interior of a *sorbent* (substrate).

The relations between the different intermolecular interactions that are responsible for sorption in a true dyeing system are incompletely known, since it is difficult to differentiate experimentally between the different interacting forces.

Electrostatic forces. The fact that, when dyeing wool, silk and polyamides with anionic dyes, and dyeing polyacrylonitrile and polyester fibres containing sulpho groups with cationic dyes, there is a stoichiometric relation between the ionic charge of the dye and the number of the gegenion charges on the fibre, demonstrates qualitatively that ionic interactions participate (ion exchange, which can be formally described by the Langmuir isotherm). This Coulomb attraction and other electrostatic forces (e.g. dipole–dipole interaction) generally diminish as the inverse second power of the distance between the charge centres.

Van der Waals' forces. Quantum mechanical considerations demonstrate[7] that two chemically-inert molecules attract one another with the so-called London force (one of the van der Waals' forces) whereby the energy of attraction diminishes as the inverse sixth power of the distance. At very small distances the repulsion of the atomic nuclei also becomes noticeable and must be taken into account in the total energy as a quantity which is inversely proportional to the twelfth power of the distance (Lennard-Jones' intermolecular potential: cf. also reference 7).

‡ The dyeing process is spontaneous ($dS/dt > 0$) and, though in practice it may be 'chemically' reversible, it is irreversible in the thermodynamic sense; it is thermodynamically reversible only in the extreme case of macroscopic equilibrium ($dS/dt = 0$) (cf. reference 6).

The assumption that such van der Waals forces bind the dye to the fibre therefore presupposes that as many dye molecules as sterically possible can approach the fibre molecules. With dyes which have an affinity for cellulose this is favoured by the planar structure of the dye ions and the inner surface of cellulose (cf., e.g., direct dyes, section 4.2.5, and the leuco form of vat dyes, section 10.5.5).

Hydrogen bonds. Qualitative relations can be demonstrated between the sorption energy of dyes and their capacity for forming intermolecular hydrogen bonds (cf. also section 10.2.1). As a result, the OH group of Orange I (**4.1**), for instance, forms an intermolecular H bond with the oxygen atom of amide groups of wool,[19] whilst in Orange II (**4.2**) there is only one intramolecular bond with the azo nitrogen present.

Hydrophobic interactions.[8] The state of equilibrium of a system composed of different types of molecules is determined not only by interaction enthalpies which result from the different sorptive forces but also by the entropy corresponding to this state. The model of Kauzmann, Scheraga and other authors assumes that hydrophobic molecules or parts of molecules which are dissolved in water (e.g. aliphatic side chains of dyes: cf. disperse dyes, sections 4.2.2 and 10.5.2) cause an ice-like structure of the water molecules to develop in the immediate vicinity. The system evades this process, which is unfavourable as regards entropy, by the association of hydrophobic molecules with one another (soap micelles, dye aggregates) or with the fibre. However, definite statements are not yet possible as our knowledge on the structure of water is insufficient.[8]

Most of the experimentally-determined dye isotherms are classified in the colouristic literature (Figure 11.2) as Langmuir, Freundlich or Nernst† isotherms, since they are functionally identical with these isotherms, which properly apply to gases.

Kinetic derivation of dye isotherms. Nomenclature:

c_{mi}^{σ}‡ = concentration of the freely mobile molecules of the dye i in the dyeing bath (σ).

c_{mi}^{φ}‡ = concentration of the mobile molecules of the dye i in the liquid in the pores of the substrate (φ).

c_{ai}^{φ} = concentration of the adsorbed molecules of the dye i.

s_j^{φ} = total concentration of adsorption sites j having the same adsorption energy E_j.

$K_{ji} = k_{ji}/k_{-ji}$ = adsorption constant of the dye i for the adsorption sites j; k_{ji} and k_{-ji} are the adsorption and desorption rate constants, respectively.

$\Theta_{ji} = c_{ai}^{\varphi}/s_j^{\varphi}$ = fractional coverage of the dye i on the adsorption sites j.

x_i = mole fraction of the substance i.

† In the literature it is customary to describe all rectilinear distribution isotherms as Nernst isotherms. Here it should be noted that the choice of different methods of expressing concentration (e.g. mole fraction instead of molarity) can functionally convert a Nernst isotherm into a Langmuir isotherm (and vice versa).[9] As a matter of definition only a Nernst isotherm with concentrations expressed as mole fractions is termed an *ideal* distribution isotherm.

‡ Since, generally speaking, the concentration c_{mi}^{φ} cannot be ascertained when evaluating experimental results, it is replaced by c_{mi}^{σ} with the tacit assumption of the rectilinear relation (11.4).

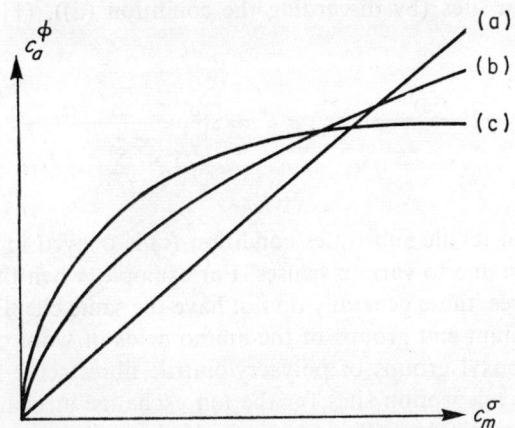

Figure 11.2 Isotherms for dyeing equilibria; (a) Nernst, (b) Freundlich, (c) Langmuir

For the simplest model it is initially assumed that
(a) the dye molecules attach themselves to the pore surfaces of a textile substrate as monomolecular particles,
(b) there are no interactions between the adsorbed dye molecules,
(c) the same adsorption energy is liberated at each adsorption site of the surface,
(d) only *one* dye competes as sorbate for the adsorption sites.§

For the sorption rate (v_1) and the desorption rate (v_{-1}), (11.1) and (11.2) are valid.

$$v_1 = k_{11} c_{m1}^{\varphi} (s_j^{\varphi} - c_{a1}^{\varphi}) \tag{11.1}$$

$$v_{-1} = k_{-11} c_{a1}^{\varphi} \tag{11.2}$$

At equilibrium $v_1 = v_{-1}$; therefore, as Langmuir has shown, the isotherm (11.3) follows.

$$c_{a1}^{\varphi} = s_1^{\varphi} \frac{c_{m1}^{\varphi} K_{11}}{1 + c_{m1}^{\varphi} K_{11}} \quad \text{or} \quad \Theta_1 = \frac{c_{m1}^{\varphi} K_{11}}{1 + c_{m1}^{\varphi} K_{11}} \quad \text{or} \quad K_{11} c_{m1}^{\varphi} = \frac{\Theta_1}{1 - \Theta_1} \tag{11.3}$$

So far, all attempts to measure c_{mi}^{φ} have been unsuccessful. In the evaluation of experimental data, therefore, the rectilinear relation (11.4) is always tacitly assumed; this does not cause any change in the functional form of (11.3) since c_{m1}^{φ} is replaced by c_{m1}^{σ} and K_{11} by $K_{11} n_1 = K_{11}'$. If we elaborate the above model and allow more than one (e.g. $i = n$) sorbate to compete

$$c_{mi}^{\varphi} = n_i c_{mi}^{\sigma} \tag{11.4}$$

§ This case is hypothetical since at least the solvent (e.g. water) competes with the dye for the adsorption sites. This fact is later taken into account in equation (11.5).

for the adsorption sites (by discarding the condition (d)), (11.3) can readily be generalized as:

$$c_{a1}{}^{\varphi} = s_1{}^{\varphi} \frac{c_{m1}{}^{\varphi} K_{11}}{1 + \sum_{i=1}^{n} c_{mi}{}^{\varphi} K_{1i}} \quad \text{or} \quad c_{a2}{}^{\varphi} = s_1{}^{\varphi} \frac{c_{m2}{}^{\varphi} K_{12}}{1 + \sum_{i=1}^{n} c_{mi}{}^{\varphi} K_{1i}} \quad \text{and so on} \quad (11.5)$$

With the dyeing of textile substrates condition (c) is obeyed in only the rarest of cases. This may be due to various causes. For example, when the adsorption sites are localized centres, these generally do not have the same chemical structure. The numerous ammonium end groups of the amino acids of wool or the sulphonato, sulphato and carboxyl groups of polyacrylonitrile fibres serve as examples since they represent the adsorption sites for the ion exchange mechanism when dyeing with anionic or cationic dyes. This fact, in itself, demands refinement of the above model. Even with chemically uniform adsorption centres (e.g. with polyamides) their accessibility varies, and, consequently, the liberated adsorption energy varies (because it is inversely proportional to the distance between sorbate and sorbent), as schematically depicted in Figure 11.4(b) for the dye ⊖ As a further generalization of (11.5), there will be for q different types of adsorption sites, which are in themselves identical, a sum of q different Langmuir isotherms (11.6). As has been demonstrated on different occasions,[10] in certain conditions a sum of Langmuir isotherms can be formally approximated by means of a Freundlich isotherm (for one sorbate: (11.7)).

$$c_{a1}{}^{\varphi} = \sum_{j=1}^{q} s_j{}^{\varphi} \frac{c_{m1}{}^{\varphi} K_{j1}}{1 + \sum_{i=1}^{n} c_{mi}{}^{\varphi} K_{ji}} \quad \text{or} \quad c_{a2}{}^{\varphi} = \sum_{j=1}^{q} s_j{}^{\varphi} \frac{c_{m2}{}^{\varphi} K_{j2}}{1 + \sum_{i=1}^{n} c_{mi}{}^{\varphi} K_{ji}} \quad \text{and so on} \quad (11.6)$$

$$c_{a1}{}^{\varphi} = K(c_{m1}{}^{\varphi})^z \quad (z < 1) \quad (11.7)$$

However, experimental data that can be described by means of a Freundlich isotherm can also be explained by numerous other dyeing mechanisms. For instance, taking into account the marked aggregation tendency of the mobile dye molecules in the dye bath (σ) and/or in the pores of the substrate (φ) results in the expression (11.8) for the total concentration of the free dye 1 in the pore liquid. Here, n denotes the degree of aggregation. If the dye is almost exclusively present as the dimer (that is to say, the relation (11.10) is valid) and only the monomeric dye is adsorbed, it results, in accordance with (11.2) and taking into consideration (11.8) and (11.9), in the isotherm (11.11) which for $1 > K_{11}'\sqrt{c_{m1}{}^{\varphi}}$ is transformed into the Freundlich isotherm ((11.7): $z = 0.5$). When the dye consists almost exclusively of monomer but, in the main, dimer is adsorbed, a Freundlich isotherm results (11.7), with $z = 2$ for low dye concentrations.

$$(c_{m1}^{\varphi})_{\text{tot}} = \sum_{n=1}^{\infty} n(c_{m1}^{\varphi})_n \qquad (11.8)$$

$$(c_{m1}^{\varphi})_{n=2} = (c_{m1}^{\varphi})_{n=1}^{2} K_D \qquad (11.9)$$

$$(c_{m1}^{\varphi})_{n=1} \approx \sqrt{\frac{(c_{m1}^{\varphi})_{\text{tot}}}{2K_D}} = \sqrt{\frac{c_{m1}^{\varphi}}{2K_D}}, \quad \text{when } (c_{m1}^{\varphi})_{n=2} > (c_{m1}^{\varphi})_{n \neq 2} \quad (11.10)$$

$$c_{a1}^{\varphi} = s_1^{\varphi} \frac{K_{11}\sqrt{c_{m1}^{\varphi}/2K_D}}{1 + K_{11}\sqrt{c_{m1}^{\varphi}/2K_D}} = s_1^{\varphi} \frac{K_{11}'\sqrt{c_{m1}^{\varphi}}}{1 + K_{11}'\sqrt{c_{m1}^{\varphi}}} \qquad (11.11)\dagger$$

The generalization of our model invalidates condition (a) by assuming the adsorption of dimeric (or polymeric) dye and therefore causes the fractional coverage Θ_{ji}, which is defined as $c_{ai}^{\varphi}/s_j^{\varphi}$, to exceed the limiting value of 1 which can be

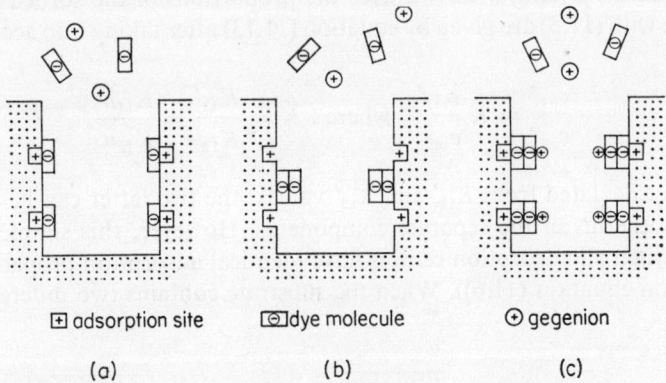

Figure 11.3 Schematic representation of the dyeing process by the ion-exchange mechanism

attained in monomolecular adsorption. Thus this model explains so-called 'over dyeing', a phenomenon which frequently occurs when dyeing polyamides with dyes which readily aggregate.[11,12,13] The following argument helps to explain this interesting effect: Figure 11.3(a) depicts schematically a pore of a substrate which is dyed in accordance with the stoichiometric ion exchange mechanism with positively charged adsorption sites (e.g. ammonium end groups). This dyeing fulfils the condition (b) of our model. However, a dye having a strong aggregation tendency will endeavour to attain the state sketched in Figure 11.3(b). Although this state no longer fulfils condition (b) it still obeys the stoichiometric relation. It can have either less energy or more energy than that depicted in Figure 11.3(a), depending on the amount of energy resulting from the gain of entropy in aggregation as

† When c_{mi}^{φ} is very small (11.9) no longer applies and (11.11) is transformed into (11.3).

168   *Fundamentals of the Chemistry and Application of Dyes*

compared with the energy required for the separation of the opposing charges.‡ With regard to 'overdyeing', it is of decisive importance that the state sketched in Figure 11.3(c) has lower energy than the two others in Figures 11.3(a) and 11.3(b) even though additional positive charge must be introduced into the matrix which is already positive.§

The so-called Nernst isotherm ‖ (11.12) can be regarded as a limiting case of the Langmuir isotherm ((11.5): $1 \gg \sum_{i=1}^{n} c_{mi}^{\varphi} K_{1i}$) or of the Freundlich isotherm ((11.7): $z = 1$).

$$c_{a1}^{\varphi} = K c_{m1}^{\varphi} \qquad (11.12)$$

The equilibrium behaviour of dyeing systems with more than one sorbate (for instance mixed dyes or dye solutions containing salt or carriers) can also generally be interpreted by means of our simple model with the aid of equation (11.5): e.g. in a combination dyeing with two dyes the proportions of the sorbed amounts in accordance with (11.5) are given by equation (11.13) after taking into account (11.4).

$$\frac{c_{a1}^{\varphi}}{c_{a2}^{\varphi}} = K \frac{c_{m1}^{\sigma}}{c_{m2}^{\sigma}} \quad \text{where} \quad K = \frac{K_{11}'}{K_{12}'} = \frac{K_{11} n_1}{K_{12} n_2} \qquad (11.13)$$

K can be calculated from K_{11}' and K_{12}' values; the two latter can be determined from the isotherms of the separate components. However, this simple relation is valid only when all adsorption centres have identical adsorption properties (cf. the discussion on equation (11.6)). When the substrate contains two different types of

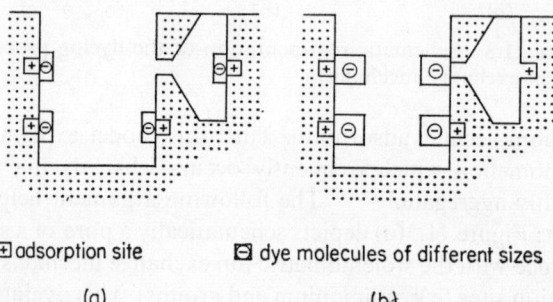

⊞ adsorption site   ⊟ dye molecules of different sizes

(a)   (b)

Figure 11.4 Schematic diagram of sterically-selective adsorption

‡ It can be calculated that the most energetically favourable location of the dimer can be, but need not always be, half-way between the two positive charges (Figure 11.3(b)).
§ This process can have a high activation energy.
‖ Cf. first footnote on page 164.

adsorption sites equations (11.14)–(11.16) apply, provided that no sterically selective adsorption (cf. Figure 11.4(b)) occurs.

$$\frac{c_{a1}^{\varphi}}{c_{a2}^{\varphi}} = \frac{s_1^{\varphi}\dfrac{c_{m1}^{\varphi}K_{11}}{1+c_{m1}^{\varphi}K_{11}+c_{m2}^{\varphi}K_{12}} + s_2^{\varphi}\dfrac{c_{m1}^{\varphi}K_{21}}{1+c_{m1}^{\varphi}K_{21}+c_{m2}^{\varphi}K_{22}}}{s_1^{\varphi}\dfrac{c_{m2}^{\varphi}K_{12}}{1+c_{m1}^{\varphi}K_{11}+c_{m2}^{\varphi}K_{12}} + s_2^{\varphi}\dfrac{c_{m2}^{\varphi}K_{22}}{1+c_{m1}^{\varphi}K_{21}+c_{m2}^{\varphi}K_{22}}} \quad (11.14)$$

Taking into account (11.4) and provided that $m_1 = n_2K_{12}/n_1K_{11}$, $m_2 = K_{21}/K_{11}$, $m_3 = K_{22}/K_{12}$, we obtain:

$$\text{For} \quad c_{m1}^{\varphi} \to \text{large:} \quad \frac{c_{a1}^{\varphi}}{c_{a2}^{\varphi}} = \frac{s_1^{\varphi}+s_2^{\varphi}}{m_1(s_1^{\varphi}+s_2^{\varphi}m_3/m_2)} \cdot \frac{c_{m1}^{\sigma}}{c_{m2}^{\sigma}} \quad (11.15)$$

$$\text{For} \quad c_{m2}^{\varphi} \to \text{large:} \quad \frac{c_{a1}^{\varphi}}{c_{a2}^{\varphi}} = \frac{s_1^{\varphi}+s_2^{\varphi}m_2/m_3}{m_1(s_1^{\varphi}+s_2^{\varphi})} \cdot \frac{c_{m1}^{\sigma}}{c_{m2}^{\sigma}} \quad (11.16)$$

Sterically selective adsorption occurs when certain adsorption sites, which are otherwise identical, are inaccessible to one adsorbate for steric reasons (cf. Figure 11.4(b)). Therefore, from the point of view of adsorption energy, only one type of adsorption centre exists for compound $\boxed{\ominus}$ but there are several for $\boxed{\ominus}$.

Straight-line isotherms† are always used for the evaluation of experimental results.

$$c_{a1}^{\varphi} = K(c_{m1}^{\varphi})^z \qquad \log c_{a1}^{\varphi} = z \log c_{m1}^{\varphi} + \log K \quad (11.17)$$

$z = 1$: Nernst, rectilinear $\qquad$ Gradient $z = 1$: Nernst, rectilinear

$z \neq 1$: Freundlich, curved $\qquad$ Gradient $z \neq 1$: Freundlich, rectilinear

The Langmuir isotherm can be converted to this form in different ways by the transformations (11.18)–(11.21); (11.21) (a modification of the Scatchard presentation (11.20)) is particularly suitable for the investigation of competitive dyeing processes (mixtures of dyes, dyes with salt or carrier additions, etc.).

$$\frac{1}{c_{a1}^{\varphi}} = \frac{1}{c_{m1}^{\varphi}} \frac{1}{K_{11}s_1^{\varphi}} \left[1 + \sum_{i=2}^{n} c_{mi}^{\varphi}K_{1i}\right] + \frac{1}{s_1^{\varphi}} \quad (11.18)$$

$$\frac{c_{m1}^{\varphi}}{c_{a1}^{\varphi}} = \frac{c_{m1}^{\varphi}}{s_1^{\varphi}} + \frac{1}{K_{11}s_1^{\varphi}}\left[1 + \sum_{i=2}^{n} c_{mi}^{\varphi}K_{1i}\right] \quad (11.19)$$

$$\frac{c_{a1}^{\varphi}}{c_{m1}^{\varphi}}\left[1 + \sum_{i=2}^{n} c_{mi}^{\varphi}K_{1i}\right] = -K_{11}c_{a1}^{\varphi} + K_{11}s_1^{\varphi} \quad (11.20)$$

$$\frac{c_{a1}^{\varphi}}{c_{m1}^{\varphi}} = -K_{11}\left[c_{a1}^{\varphi} + \sum_{i=2}^{n} c_{ai}^{\varphi}\right] + K_{11}s_1^{\varphi} \quad (11.21)$$

Finally, some of the mathematical expressions for the Langmuir isotherm that are frequently used in the trade literature will be compared with one another;‡ isotherm

---

† The second footnote on page 164 for the following equations (11.17)–(11.25) should be noted.

‡ In the ion-exchange mechanism on which our comparison is based at least two substances compete for the adsorption sites, on account of electrical neutrality.

(11.22), which is frequently regarded as valid for the so-called ion-exchange mechanism, as well as the much-used expression (11.23) (x = molar fraction of the dye), derived with the aid of the Donnan membrane theory, yield the 'adsorption constants' K' and K''; these are not identical with the constant K_{11} of our model (11.25).

$$\frac{\Theta_1}{1-\Theta_1} = K'c_{m1}^\varphi; \qquad \Theta_1 = \frac{c_{a1}^\varphi}{s_1^\varphi} \qquad (11.22)$$

$$\frac{\Theta_1}{1-\Theta_1} = K'' \frac{x_1^\varphi}{1-x_1^\varphi} = K'' \frac{c_{m1}^\varphi}{c_{m2}^\varphi} \qquad (11.23)$$

(11.5) establishes that:

$$\frac{\Theta_1}{1-\Theta_1} = \frac{K_{11}}{1+K_{12}c_{m2}^\varphi} c_{m1}^\varphi \qquad (11.24)$$

Thus:

$$K' = \frac{K_{11}}{1+K_{12}c_{m2}^\varphi}; \qquad K'' = \frac{K_{11}c_{m2}^\varphi}{1+K_{12}c_{m2}^\varphi} \qquad (11.25)$$

For $c_{m1}^\varphi \ll c_{a1}^\varphi$ and $c_{m2}^\varphi \ll c_{a2}^\varphi$, then $K' = K_{11}/K_{12}c_{m2}^\varphi$ and $K'' = K_{11}/K_{12}$.

This clearly shows how little confidence can be placed in the standard affinities ($RT \ln K$) reported by various authors. In contrast to K_{11} (or K_{12}), the K' and K'' values directly depend on the presence of other substances (salts, other dyes, etc.) (cf. (11.25)).

Thermodynamic description of the dyeing equilibria. When ∂n mol/l are added to a phase containing n mol/l of a dye and all other variables being kept constant, the free energy will change by ∂G; thus the chemical potential μ of this dye is defined by the relation (11.26); this is exponentially related (11.27) to the absolute activity λ of the dye and it allows a particularly clear presentation of the laws of mixtures.

$$\frac{\partial G}{\partial n} = \mu \qquad (11.26)$$

$$\lambda = \exp(\mu/RT) \qquad (11.27)$$

λ is a function of the concentrations of the components of a mixture, and for component 1 of a binary mixture† it is, for example, generally given by (11.28); here n_1 and n_2 denote the number of moles of the components 1 and 2; and x_1 is the molar fraction relating to component 1; $\Phi(x_1)$ means a function of x_1, and is called the Hildebrand activity coefficient. The proportionality factor λ_1^0 indicates the standard value of the absolute activity; it can be obtained from (11.28) for

$$\lambda_1 = \lambda_1^0 \frac{n_1}{n_1+n_2} \Phi\left(\frac{n_1}{n_1+n_2}\right) = \lambda_1^0 x_1 \Phi(x_1) \qquad (11.28)$$

† If there is a large excess of one component over the other this mixture is usually described as a solution.

$x_1 = 1$ and $\Phi(x_1) = 1$ (pure component). Where $\Phi(x_1) = 1$ over the entire concentration range a rectilinear relation results from (11.28) which is known as the Raoult law; a mixture which can be described by means of the Raoult law is termed 'ideal'. When $\Phi(x_1)$ does not obey this condition, (11.28) describes a non-ideal mixture. In respect of $\Phi(x_1)$ two limiting cases can be assumed as reference states for λ_1, namely the pure component 1 (11.29) or the infinitely dilute mixture (solution) (11.30).

$$\lim_{x_1 \to 1} \Phi(x_1) = 1 \tag{11.29}$$

$$\lim_{x_1 \to 0} \Phi(x_1) = h_1 \tag{11.30}$$

In the first case, λ_1^0 represents the standard value for λ_1, while in the second case it is usual to choose $\lambda_1^\ominus = \lambda_1^0 h_1$ as the new standard value for λ_1—thus (11.28) is replaced by the equivalent relation (11.31) ($\Phi(x_1) = h_1 \psi(x_1)$). When $\psi(x_1)$, the so-called Lewis activity coefficient, equals 1, (11.31) is transformed into a rectilinear relation which is known as Henry's law. In the case of solutions of a solid in a liquid (e.g. dyes) it is also possible to choose the saturated solution as the reference state. By analogy with equations (11.28) and (11.31), relation (11.32) now becomes valid (L_1 denotes the solubility of component 1 expressed as mole fraction), and hence (11.33), for ideally dilute solutions, is derived.

$$\lambda_1 = \lambda_1^0 h_1 x_1 \Psi(x_1) = \lambda_1^\ominus x_1 \Psi(x_1); \qquad \lambda_1^0 h_1 = \lambda_1^\ominus \tag{11.31}$$

$$\lambda_1 = \frac{\lambda_1^0}{L_1} x_1 \theta(x_1) = \lambda_1^\bullet x_1 \theta(x_1) \tag{11.32}$$

For:
$$\lim_{x_1 \to 0} \theta(x_1) = h_1 L_1: \qquad \lambda_1 = \lambda_1^\bullet h_1 L_1 x_1 \tag{11.33}$$

It is possible to define further reference systems in addition to the three mentioned above. The system chosen is a matter of convention and is largely decided by the experimentally accessible data.

For a dyeing process the chemical potential of a dye (component F) in the dye bath (σ) or in the substrate (φ) can be described by means of (11.34) or (11.35) with the aid of (11.27) and (11.28).† When there are no chemical reactions, and no aggregates are formed, this gives the relation (11.36) for the sorption process ($\Delta \mu_F$ or $\Delta \mu_F^0$ is written instead of $\mu_F^\varphi - \mu_F^\sigma$ or $\mu_F^{0\varphi} - \mu_F^{0\sigma}$, respectively). This relation corresponds to the definition by de Donder[14] for the affinity ($\Delta \mu_F$).

---

† Analogous relations can be formulated with the aid of equations (11.27) and (11.31) or (11.32) for other reference states.

$$\mu_F^{\sigma} = \mu_F^{0\sigma} + RT \ln x_F^{\sigma}\Phi(x_F^{\sigma}) \qquad (11.34)$$

$$\mu_F^{\varphi} = \mu_F^{0\varphi} + RT \ln x_F^{\varphi}\Phi(x_F^{\varphi}) \qquad (11.35)$$

$$\Delta\mu_F = \Delta\mu_F^{0} + RT \ln \frac{x_F^{\varphi}\Phi(x_F^{\varphi})}{x_F^{\sigma}\Phi(x_F^{\sigma})} \qquad (11.36)$$

Therefore at equilibrium $\Delta\mu_F = 0$ (or $\lambda_F^{\varphi} = \lambda_F^{\sigma}$) and the standard affinity‡ for the distribution of a substance between two non-miscible phases is proportional to the logarithm of the quotient of the standard values of the absolute activities of the substance in both phases. The numerical value of the standard affinity depends not only on the selected reference state (Table 11.1) but also on the units of concentration, so that it is even possible to obtain different values for the standard affinity with the same reference state: one for mole fraction x ($\Delta\mu_F^0$, $\Delta\mu_F^{\ominus}$, $\Delta\mu_F^{\bullet}$), and another for molality m($\Delta\mu_F^{0m}$, $\Delta\mu_F^{\ominus m}$, $\Delta\mu_F^{\bullet m}$) and another for molarity m' ($\Delta\mu_F^{0m'}$, $\Delta\mu_F^{\ominus m'}$, $\Delta\mu_F^{\bullet m'}$). In principle they can all be interconverted[5,9] but with textile substrates none of the required exact data, such as the effective number of moles n^{φ} or the molecular weight M^{φ} are experimentally accessible.§ It follows that the use of absolute values of standard affinities for dyeing processes is rather dubious. At best their application could possibly be meaningful for the purposes of comparison in one and the same dyebath-substrate system.

Table 11.1 Standard affinities and their reference states

Reference state	Standard affinity	Equation
Pure substance	$\Delta\mu_F^0 = RT \ln \dfrac{\lambda_F^{0\sigma}}{\lambda_F^{0\varphi}} = RT \ln \dfrac{x_F^{\varphi}\Phi(x_F^{\varphi})}{x_F^{\sigma}\Phi(x_F^{\sigma})} = 0$	(11.37)
Infinitely dilute solution	$\Delta\mu_F^{\ominus} = RT \ln \dfrac{h_F^{\sigma}}{h_F^{\varphi}} = RT \ln \dfrac{x_F^{\varphi}\Psi(x_F^{\varphi})}{x_F^{\sigma}\Psi(x_F^{\sigma})}$	(11.38)
Saturated solution	$\Delta\mu_F^{\bullet} = RT \ln \dfrac{L_F^{\varphi}}{L_F^{\sigma}} = RT \ln \dfrac{x_F^{\varphi}\theta(x_F^{\varphi})}{x_F^{\sigma}\theta(x_F^{\sigma})}$	(11.39)

Dyeing isotherms can also be derived on the basis of these thermodynamic considerations; however, care must be taken that an exact derivation proceeds from equations which have dimensional agreement.† The Nernst isotherm‖ (11.40)

‡ In the literature, frequently denoted as $\Delta\mu_F^0$ or as $-\Delta\mu_F^0$ without indication of the reference state.
§ It has not been decided whether the lengths of whole fibre molecules or the lengths of the mobile segments should be used to calculate the mole fraction x.
† In numerous publications and monographs[15,16] this condition is disregarded. Therefore, the equations for the standard affinity of a dyeing system, for example, which are based on Gilbert and Rideal's[17] theory have no thermodynamic foundation, and although the theory can fit many experimental results, it is limited by approximations, for example the use of equation (11.22) for (11.47).
‖ Compare first footnote page 164.

follows directly from (11.38) or (11.39) when the ψ or θ function equals 1. The Langmuir isotherm can be derived similarly from (11.38) (cf. (11.42), (11.43)). For instance, the ion exchange mechanism can serve as a model (11.41). It follows from the partition functions for a perfect solution with a limited number of voids (sorption sites) that $\psi(x_F) = 1/(1 - x_F)$, where x_F signifies the fractional coverage of the dye (for a substrate this is also written as Θ_F). When the infinitely dilute solution is chosen as reference state the standard state is $x_F = 0.5$ (cf. 11.31).

$$x_F{}^\varphi = \frac{h_F{}^\sigma}{h_F{}^\varphi} x_F{}^\sigma = \frac{L_F{}^\varphi}{L_F{}^\sigma} x_F{}^\sigma = K x_F{}^\sigma \tag{11.40}$$

$$(\text{Dye})^\sigma + (\text{competing ion})^\varphi \rightleftarrows (\text{Dye})^\varphi + (\text{competing ion})^\sigma \tag{11.41}$$

$$\mu_F{}^\varphi = \mu_F{}^{0\varphi} + RT \ln h_F{}^\varphi + RT \ln x_F{}^\varphi - RT \ln (1 - x_F{}^\varphi) \tag{11.42}$$

$$\mu_F{}^\sigma = \mu_F{}^{0\sigma} + RT \ln h_F{}^\sigma + RT \ln x_F{}^\sigma - RT \ln (1 - x_F{}^\sigma) \tag{11.43}$$

Index F = dyestuff; analogous relations also apply to the competing ion.

In accordance with (11.38) the equation (11.45) results from the equilibrium condition (11.44)§ at the standard state ($\Delta \mu_F{}^0 = 0$). For the sake of simplicity it is assumed that the gegenion is not taken up by the substrate (cf. Figure 11.3(a)), so that the number of voids (sorption sites) remains constant in both phases. A relation which is analogous to equation (11.23) ($x_1{}^\varphi = x_F{}^\sigma$, $\Theta_1 = x_F{}^\varphi$) follows from (11.46).

$$\Delta \mu_F = \mu_F{}^\varphi - \mu_F{}^\sigma = 0 \tag{11.44}$$

$$\Delta \mu_F{}^\ominus = RT \ln \frac{h_F{}^\sigma}{h_F{}^\varphi} = RT \ln \frac{x_F{}^\varphi(1 - x_F{}^\sigma)}{x_F{}^\sigma(1 - x_F{}^\varphi)} \tag{11.45}$$

$$\frac{h_F{}^\sigma}{h_F{}^\varphi} = K = \frac{x_F{}^\varphi(1 - x_F{}^\sigma)}{x_F{}^\sigma(1 - x_F{}^\varphi)} \; ; \quad \frac{x_F{}^\varphi}{1 - x_F{}^\varphi} = K \frac{x_F{}^\sigma}{1 - x_F{}^\sigma} \tag{11.46}$$

$\psi(x_F{}^\sigma) = 1$ applies to the adsorption of non-ionic dyes on a textile substrate with a limited number of equivalent adsorption sites and this leads to the relation (11.47). In this case $x_F{}^\sigma = c_{m1}{}^\sigma/(c_L{}^\sigma + c_{m1}{}^\sigma)$, $c_L{}^\sigma$ being the concentration of the water (or the solvent). Since $c_L{}^\sigma \gg c_{m1}{}^\sigma$, $c_L{}^\sigma$ which is practically constant, can be included in the constant; where (11.4) applies equation (11.22) results.

$$K x_F{}^\sigma = \frac{\Theta_F}{1 - \Theta_F} \tag{11.47}$$

More complicated dyeing systems can be treated analogously; in this way equilibrium relations are frequently obtained that can be described (to a first

---

§ $\Delta G^\ominus$ is defined as $\mu^{\ominus\sigma} - \mu^{\ominus\varphi}$, whilst from (11.38) $\Delta \mu^\ominus = \mu^{\ominus\varphi} - \mu^{\ominus\sigma}$; hence $\Delta G^\ominus = -\Delta \mu^\ominus$.

approximation) by means of Freundlich isotherms (11.7). The corresponding standard dyeing entropy (ΔS^0; $\Delta S^\ominus$; $\Delta S^\bullet$) can be calculated from the standard affinity ($\Delta \mu^0$; $\Delta \mu^\ominus$; $\Delta \mu^\bullet$) and from the standard heat of dyeing (ΔH^0; $\Delta H^\ominus$; $\Delta H^\bullet$) with the aid of the Gibbs–Helmholtz equation ((11.48): $\ominus$ as reference state).

$$-\Delta \mu^\ominus = \Delta G^\ominus = \Delta H^\ominus - T \Delta S^\ominus \qquad (11.48)$$

$$\Delta H^\ominus = d\left(\frac{-\Delta \mu^\ominus}{T}\right) \Big/ d\left(\frac{1}{T}\right) \qquad (11.49)$$

It is possible to determine the standard heat of dyeing experimentally as a tangent to the function $-\Delta \mu^\ominus / T = f(1/T)$ (11.49). Since this is always negative, an increase in temperature shifts the dyeing equilibrium in the direction of higher dyebath concentration, in accordance with (11.48). $\Delta H^\ominus$, as a measure of the binding energy, can give information regarding the type of bond between dye and substrate.[18,19] The evaluation of standard dyeing entropies ($\Delta S^\ominus$) is still in the early stages.[8]

11.3.2. The kinetics of dyeing

The kinetic behaviour of a dye in its transfer from the dye bath on to or into the substrate is graphically depicted by the so-called rate of dyeing curve (cf. Figure 11.1). This is composed of numerous time-dependent processes (e.g. diffusion, adsorption, chemical reaction) and only in rare cases can these be exactly formulated.

Like dyeing equilibria, dyeing kinetics can be discussed from a thermodynamic point of view. This can be done with the aid of the thermodynamics of irreversible processes.[20] Such a thermodynamic phenomenological description of the dyeing process makes a rigorous solution of the problem of dyestuff transfer in textile substrates theoretically possible, but leads to a formula which is ill-suited for the evaluation of experimental results. Despite such reservations, this macroscopic approach to the study of dyeing can give investigators useful hints concerning physically meaningful variables for the closer examination of dyeing phenomena.

Statistical-mechanical models which are based on a kinetic interpretation of molecular processes give a clearer picture of the dyeing process.

In the following we shall demonstrate for the benefit of dye chemists engaged in experimental work, by means of a simple easily-understood model, some features of dyeing mechanisms that are indispensable for the interpretation of results in terms of molecular phenomena.

Figure 11.5 schematically depicts a pore of a textile substrate that is to be dyed. Assuming that there is good circulation, i.e. sufficiently rapid mass transfer in the dyeing bath, the dyeing process in general consists of the following processes:
(a) diffusion of ① to the substrate,
(b) immobilization (fixation) of ① (e.g. adsorption, chemical reaction),
(c) liberation of ② (e.g. desorption),
(d) diffusion of ② out of the substrate.

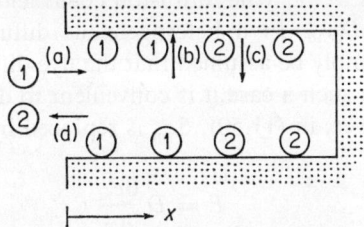

Figure 11.5 Schematic representation of a pore in textile substrates. ① and ② different sorbate molecules

The diffusion of the mobile substances ① or ② (process (a) or (d)) can be regarded as a 'random walk' and can therefore be described by the rate equation (11.50) (Fick's 1st law). Accordingly, the mass flow per unit area (flux) F is proportional to the concentration gradient dc_m/dx of the mobile molecules. The proportionality factor D is termed 'diffusion coefficient' (or 'diffusivity') and is a measure of the facility with which the free molecules can move in the appropriate medium; D is therefore similar to conductivity, a property of the diffusion medium, and it is greater the greater the mean free path of the diffusing molecules. The mean free path is determined by the total concentration of all molecules that are present.† Thus, for the diffusion of a substance in a liquid, D is constant because the concentration of the liquid is almost constant. Figure 11.6 shows the orders of magnitude of diffusion constants in different media. For example, in aqueous dyeing systems the values range from 0.5×10^{-5} to 0.5×10^{-6} cm²/s. In a homogeneous

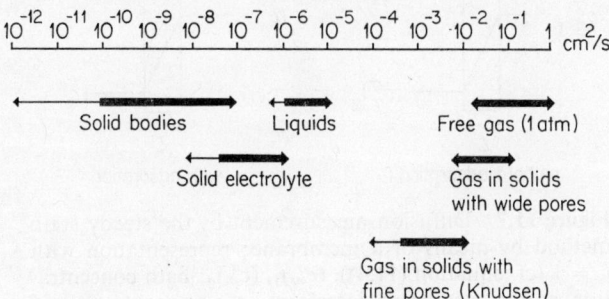

Figure 11.6 Orders of magnitude of 'true' diffusion coefficients (D_{true})

† A gas under a total pressure of 1 atm possesses a smaller diffusion coefficient than the same gas under 0·1 atm total pressure. Only in this sense is it justifiable to speak of a diffusion coefficient that is dependent on concentration.

medium D corresponds to the 'true' diffusion coefficient and in the following discussion will be called D_{true}. In a heterogeneous diffusion medium, e.g. textile fibres, *initially* it will simply be assumed that the dye diffuses into the pores, which are filled with water. In such a case it is convenient to define an effective diffusion coefficient D_{eff}; moreover, in (11.50), dc_m is replaced by dc_m^φ.

$$F = D \frac{dc_m}{dx} \quad (11.50)$$

$$D_{\text{eff}} = D_{\text{true}} \frac{P}{b} \quad (11.51)$$

P = porosity
$b = \sqrt{3} = 1\cdot7$ = tortuosity factor†

In principle, diffusion coefficients can be measured in two ways, namely under stationary or non-stationary conditions. The *steady state method* is based on Fick's 1st law (11.50); it always gives D_{eff}, and hence, in accordance with (11.51), D_{true}, irrespective of whether immobilization reactions (as, for instance, adsorption) accompany the diffusion process (cf. Figure 11.7). It requires, however, that the heterogeneous diffusion medium (e.g. textile substrate) be available in a suitable form—usually a film—for the application of (11.50). When taking into account (11.4), and then assuming stationary conditions, dc_m can be replaced by $n[(c_m^\sigma)_1 - (c_m^\sigma)_2]$ and dx by L in equation (11.50) when $(c_m^\sigma)_1$ and $(c_m^\sigma)_2$ denote the bath concentrations measured at the film surfaces and L is the thickness of the film (cf. Figure 11.7). When n‡ and P are known, D_{true} can be calculated from the mass flux F.

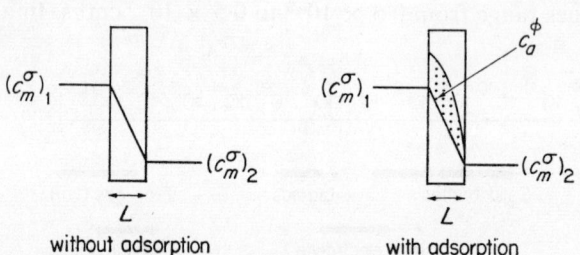

Figure 11.7 Diffusion measurement by the steady state method by means of a membrane; representation with $n = 1$ (cf. equation (11.4)). $(c_m^\sigma)_1$, $(c_m^\sigma)_2$: bath concentration at the membrane interfaces. c_a^φ: concentration of adsorbed (immobilized) substance. L: membrane thickness

† b is a factor which takes into account the random paths of the pores (tortuosity factor).
‡ Since the determination of n is experimentally very difficult (cf. (11.68)-(11.73)), it is generally included in the diffusion coefficient or frequently tacitly assumed to be 1.

However, the dye chemist is frequently confronted with situations in which the diffusion measurements are made under *non-steady state*.

In such cases Fick's 2nd law ((11.52): in the one-dimensional case) describes the diffusion kinetics.

$$\frac{\partial F}{\partial x} = \frac{\partial c^\varphi}{\partial t} \to \frac{\partial}{\partial x}\left(D_{\text{eff}} \cdot \frac{\partial c_m^\varphi}{\partial x}\right) = \frac{\partial c^\varphi}{\partial t} \to D_{\text{eff}} \cdot \frac{\partial^2 c_m^\varphi}{\partial x^2} = \frac{\partial c^\varphi}{\partial t} \quad (11.52)\dagger$$

$$c^\varphi = Pc_m^\varphi + \Phi(c_m^\varphi, t) \quad (11.53)$$

$$D_{\text{eff}} \cdot \frac{\partial^2 c_m^\varphi}{\partial x^2} = \frac{\partial(Pc_m^\varphi + \Phi(c_m^\varphi, t))}{\partial t} = \frac{\partial(Pc_m^\varphi + \Phi(c_m^\varphi, t))}{\partial c_m^\varphi} \cdot \frac{\partial c_m^\varphi}{\partial t}$$

$$= \left[P + \frac{\partial \Phi(c_m^\varphi, t)}{\partial c_m^\varphi}\right] \cdot \frac{\partial c_m^\varphi}{\partial t} \quad (11.54)$$

$$D_{\text{app}} = D_{\text{eff}} \bigg/ \left(1 + \frac{1}{P} \cdot \frac{\partial \Phi(c_m^\varphi, t)}{\partial c_m^\varphi}\right) \cdot P = D_{\text{true}} \bigg/ \left(1 + \frac{1}{P} \cdot \frac{\partial \Phi(c_m^\varphi, t)}{\partial c_m^\varphi}\right) b \quad (11.55)$$

$$D_{\text{app}} \cdot \frac{\partial^2 c_m^\varphi}{\partial x^2} = \frac{\partial c_m^\varphi}{\partial t} \quad (11.56)$$

$$\frac{(\bar{c}^\varphi)_t}{(c^\varphi)_\infty} = f = 1 - 0.69\left[\exp\left(-5.78\frac{t}{\tau}\right) + 0.19 \exp\left(-30.5\frac{t}{\tau}\right) + \ldots\right] \quad (11.57)$$

general: $\quad \tau = \beta \cdot \tau_D{'} = R_0^2 \gamma / D_{\text{app}}$

especially for $\Phi(c_m^\varphi, t) = 0; \gamma = 1$: $\quad \tau = \beta \cdot \tau_D = R_0^2 b / D_{\text{true}}$ $\quad (11.58)$

β = geometric factor

Equation (11.52) states that the difference in the mass flows through two opposite interfaces of the volume element (left hand side of equation (11.52)) is equal to the amount of substance which has accumulated in unit time in the volume element (right hand side of equation (11.52)). Only free molecules (c_m^φ) are involved in the mass flux; the state of the accumulated molecules (i.e. free or fixed, or both) is not specified.

Equation (11.57) is the integrated form of equation (11.56) for a cylinder, which approximates the shape of textile fibres.‡ The fraction f is the ratio of the amount of dye $(c^\varphi)_t$ taken up by the substrate, at time t, to the dye concentration $(c^\varphi)_\infty$; according to the equilibrium, f can vary from 0 ($t = 0$) to 1 ($t \to \infty$). τ, which is proportional to the relaxation time τ_D§ of the diffusion process, depends on the

† For the three-dimensional case: $D_{\text{eff}} \nabla^2 c_m^\varphi = \partial c^\varphi / \partial t$.
‡ Cf., e.g., references 22, 27 for the exact derivation of equation (11.57) as well as the solutions for other geometrical forms.
§ The relaxation time is that time after which a process is practically completed: i.e. when the dimensionless response variable is decreased by a factor of e^{-1}.

radius R_0 of the cylinder (fibre), the apparent diffusion coefficient D_{app} and a factor γ which is determined by the functional form of the immobilization reaction $\Phi(c_m{}^\varphi, t)$. D_{true} can be obtained only when $\Phi(c_m{}^\varphi, t)$ and γ are known.

In order to be able to determine diffusion coefficients from sorption experiments one should endeavour to find those τ values which, when inserted into (11.57), give the best possible agreement with experimental data. For this purpose (11.57) can be approximated by the function (11.59) for small f values of 0 to 0·4 and D_{app} (and D_{true}) can be ascertained from its gradient.

$$f = \frac{a\sqrt{\gamma}}{\sqrt{\tau}} \cdot \sqrt{t} = \frac{a}{R_0} \sqrt{D_{app}} \cdot \sqrt{t} \tag{11.59}$$

for a cylinder: $a = 2·0$

A thermodynamic phenomenological treatment of the processes (a)–(d) (cf. page 174) which are important as regards dyeing kinetics shows that these processes exert a mutual influence on one another. Thus, for example, substance ①, when its true diffusion coefficient is larger than that of substance ②, will accelerate the transport of substance ② but will itself be slowed down. An assessment of D_{true} on the basis of the Stokes–Einstein relation (11.60) demonstrates that, to a first approximation, this mutual effect can be disregarded for molecular weights of 100 to 800 (typical molecular weights of dyes).† However, as (11.52)–(11.55) show, the influence of immobilization and liberation processes on the diffusion process cannot be neglected. Of the numerous reactions that can lead to immobili-

$$D_{true} = \frac{kT}{6\pi\eta r_0}; \qquad r_0 = \text{prop} \cdot \sqrt[3]{MG} \tag{11.60}$$

k = Boltzmann's constant; $\qquad \eta$ = viscosity of the medium

r_o = radius of the molecule, which is assumed to be spherical

MG = molecular weight of the diffusing molecules

zation of dye molecules, the following will be mentioned:
Processes that give rise to sorption equilibria:
1. The immobilization rate of a dye ① adsorbed on pore surfaces readily accessible by diffusion is given either by the speed of the adsorption process of the dye ① or through the desorption rate of another substance ②‡ that is to be replaced at these adsorption sites. Thus, the relaxation time of these processes is given either by τ_a (11.61) or by τ_d (11.62);§ here, k_1 denotes the rate constant of the adsorption

† This is by no means a contradiction of the strong mutual influence of apparent diffusion coefficients (D_{app}) which is observed mainly in ion exchange because these are linked principally via adsorption isotherms (cf. (11.5)). In an ion exchange an adsorption isotherm expresses the capability of a charged matrix to electrostatically bind an ion with the opposite charge.
‡ In combination dyeings, for instance, the adsorption rate of a dye ① is frequently limited (determined) by the desorption speed of another dye ②.
§ This is derived from the rate equation that is valid for Langmuir isotherms (cf. (11.1) and (11.2)).

process of dye ① and k_{-1} or k_{-2} that of the desorption of dye ① or of substance ② respectively.

$$\tau_a = \frac{1}{k_1 c_{m1}^\varphi + k_{-1}} \qquad (11.61)$$

$$\tau_d = \frac{1}{k_{-2}} \qquad (11.62)$$

2. When the adsorption sites are not directly on the pore surface but within the polymer mass (e.g. sulphonate or sulphate groups in polyacrylonitrile fibres) immobilization of the dye occurs through a secondary diffusion process (cf. Figure 11.8), whose relaxation time τ_v depends on the viscoelasticity of the polymer. Viscoelasticity has a temperature dependence which changes with the polymer's glass transition temperature (cf. reference 23 on the theory of glass transition temperatures.)

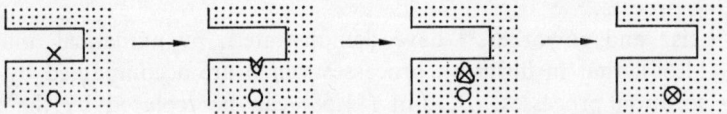

Figure 11.8 Schematic representation of the immobilization process within a polymer substance

Processes that lead to fixation by chemical reactions:

Such processes principally occur with Naphtol AS, vat and reactive dyeing processes. The fixation rate of reactive dyes on cellulose, for instance, can, as a first approximation, be described by equation (11.63). With small fractional coverage (small adsorption constant, cf. (11.3)) (11.64) applies; thus, (11.63) is transformed into (11.65) with a resulting relaxation time, τ_{fix}, which is pH dependent (11.66), due to the dissociation equilibrium of cellulose hydroxyl groups.

$$\frac{d(c_a^\varphi)_{\text{fix}}}{dt} = k c_a^\varphi c_{RO} \qquad (11.63)$$

c_{RO} = concentration of the alcoholate anion of cellulose

$$c_a^\varphi = K c_m^\varphi \qquad (11.64)$$

$$\frac{d(c_a^\varphi)_{\text{fix}}}{dt} = k K c_m^\varphi c_{RO} \qquad (11.65)$$

$$\tau_{\text{fix}} = \frac{1}{kK c_{RO}} = \frac{c_H}{kKK_{ROH} c_{ROH}} = \frac{c_H}{K'} \qquad (11.66)$$

c_{ROH} = concentration of cellulose

c_H = concentration of protons

For a more exact description of the fixation process a knowledge is needed of the cellulose anion concentration,[24] and of the influence of the preequilibria of aggregation and tautomerism on the reactivity of the dyes.[25]

The magnitude of τ_D relative to τ_a or τ_v or τ_{fix} decides which processes principally determine the dyeing kinetics:

1. $\tau_D \gg \tau_a$ or $\tau_D \gg \tau_{\text{fix}}$

Under these conditions, $\Phi(c_m{}^\varphi, t)$ can be replaced in the general (one-dimensional) formulation of Fick's 2nd law (11.54) by the isotherm values $c_a{}^\varphi$ or, in fixation reactions, by the concentration of reaction sites $(c_a{}^\varphi)_{\text{fix}}$. In the latter case, rapid chemical reactions accompany the diffusion process; this corresponds to the so-called shell progressive diffusion mechanism elaborated by P. B. Weisz and co-workers.[27] Here the adsorption constant K is infinitely large; this mechanism and pure Fickian diffusion ($K = 0$; $\Phi(c_m{}^\varphi, t) = 0$), therefore, give the two boundary conditions of all possible diffusion–sorption processes where diffusion is the rate-limiting process.[26]

P. B. Weisz and coworkers[26] have demonstrated, by numerical analysis of equation (11.54), that in diffusion processes which are accompanied by rapidly occurring sorption processes, D_{app} in (11.58) can be replaced by the relation (11.67). γ is a correction factor, which depends on the form of the adsorption isotherm, and has values ranging from 1·0 for $K = 0$ to 1·6 for $K \to \infty$;[26] n gives the distribution of dye molecules between the dyebath and the pore liquid (cf. 11.4). $c_0{}^\varphi$ and $c_0{}^\sigma$ are the constant concentrations $c_m{}^\varphi$ and $c_m{}^\sigma$, respectively, of the mobile dyestuff at the boundary of the textile substrate. $c_a{}^\varphi$ is the corresponding isotherm value for $c_0{}^\varphi$. n can be ascertained by analysis of the concentration profiles, in

$$D_{\text{app}} = \gamma D_{\text{true}} \cdot \frac{P \cdot c_0{}^\varphi}{b(P \cdot c_0{}^\varphi + c_a{}^\varphi)}$$

$$= \gamma D_{\text{true}} \cdot \frac{Pnc_0{}^\sigma}{b(Pnc_0{}^\sigma + c_a{}^\varphi)} \approx \gamma D_{\text{true}} \cdot \frac{P}{b} n \frac{c_0{}^\sigma}{c_a{}^\varphi} = \gamma \cdot Q \frac{c_0{}^\sigma}{c_a{}^\varphi} \quad (11.67)$$

$$Q = \text{permeability}$$

accordance with the 'activity gradient' model of Atherton and Peters[29] ((11.68)–(11.73)), which has been corrected by Sand.[28] By this method diffusion coefficients D_0, which are independent of the applied boundary conditions, can be determined through extrapolation of $c_m{}^\sigma \to 0$ (or $\Theta \to 0$) (cf. Figure 11.9). D_0 is connected with D_{true} by the relation (11.72).[26] Moreover, n may be calculated by comparing K with K' (adsorption constants of the 'inner' and 'outer' isotherms†), (11.73).

† Cf. discussion on equation (11.3); here: $K \equiv K_{11}$ and $K' \equiv K_{11}'$.

The Application of Dyes

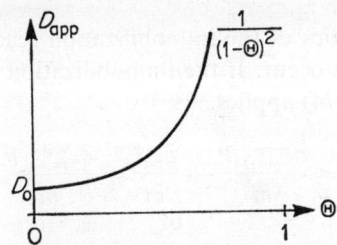

Figure 11.9 Determination of D_0 according to the activity-gradient model

For these calculations a knowledge of the porosity P is required.

$$F = D_{\text{true}} \cdot \frac{P}{b} \cdot \frac{dc_m^\varphi}{dx} = D_{\text{true}} \cdot \frac{P}{b} n \frac{dc_m^\sigma}{dx}$$

$$= D_{\text{true}} \cdot \frac{Pn}{b} \cdot \frac{dc_m^\sigma}{d(c_a^\varphi + Pc_m^\varphi)} \cdot \frac{d(c_a^\varphi + Pc_m^\varphi)}{dx} \tag{11.68}$$

$$Pc_m^\varphi + c_a^\varphi = Pc_m^\varphi + s^\varphi \frac{c_m^\varphi K}{1 + c_m^\varphi K}$$

$$= Pnc_m^\sigma + s^\varphi \frac{nc_m^\sigma K}{1 + nc_m^\sigma K} = Pnc_m^\sigma + s^\varphi \frac{c_m^\sigma K'}{1 + c_m^\sigma K'} \tag{11.69}$$

$$\frac{dc_m^\sigma}{d(Pc_m^\varphi + c_a^\varphi)} = \left[Pn + \frac{s^\varphi K'}{(1 + c_m^\sigma K')^2} \right]^{-1} = \frac{1}{Pn + s^\varphi K'(1 - \Theta)^2} \tag{11.70}$$

$$F = D_{\text{true}} \cdot \frac{Pn}{b} \frac{1}{Pn + s^\varphi K'(1 - \Theta)^2} \cdot \frac{d(c_a^\varphi + Pc_m^\varphi)}{dx}$$

$$= D_{\text{true}} \cdot \frac{P}{Pb + bs^\varphi K(1 - \Theta)^2} \cdot \frac{d(c_a^\varphi + Pc_m^\varphi)}{dx} \tag{11.71}$$

For $\Theta \to 0$ (cf. Figure 11.9):

$$D_0 = \frac{P \cdot D_{\text{true}}}{b(P + s^\varphi K)} \tag{11.72}$$

$$nK = K' \tag{11.73}$$

2. $\tau_D \ll \tau_a$ or $\tau_D \ll \tau_{\text{fix}}$ or $\tau_D \ll \tau_v$

In this extreme case the dye uptake curve is determined by the kinetics of the fixation reaction; generally speaking, the rectilinear relation (11.59) is no longer valid ('non-Fickian' diffusion: cf. reference 39).

3. $\tau_D \cong \tau_a$ or $\tau_D \cong \tau_{\text{fix}}$ or $\tau_D \cong \tau_v$

Depending on the kinetics of the immobilization reaction, deviations from the linear relation (11.59) can occur. If the immobilization reaction is first-order, the Danckwerts equation (11.74) applies.[22,37,38]

$$D_{\text{eff}} \frac{\partial^2 c_m^\varphi}{\partial x^2} = \frac{\partial c_m^\varphi \cdot P}{\partial t} + \frac{\partial c_a^\varphi}{\partial t} = \frac{\partial c_m^\varphi \cdot P}{\partial t} + k c_m^\varphi \cdot P$$

or

$$\frac{\partial c_m^\varphi}{\partial t} = \frac{D_{\text{true}}}{b} \cdot \frac{\partial^2 c_m^\varphi}{\partial x^2} - k c_m^\varphi \tag{11.74}$$

k = rate constant of the immobilization reaction

By changing boundary conditions (e.g. dye concentration in the dyebath), dyeing temperature, reactivity of the dye (in reactive dyeing) or by the addition of catalysts in order to accelerate fixation, it is possible, if desired, to approach one or other of the extreme cases (1 or 2). If, in the kinetic investigation of dyeing, extreme case 1 is found but, despite this fact, the true diffusion coefficient is significantly smaller than 10^{-6} cm²/s (cf. Figure 11.6), other phenomena must be involved: the dye–dye aggregates can markedly influence the concentration gradients,[26] or the size of the dye molecules can retard diffusion when the pores are small.[30] Moreover, it is feasible that the water within the fibre pores is partly present in a paracrystalline form (quasi-solid phase),[31] which would slow down diffusion. Additions of alcohol, urea, etc. can destroy the dye–dye aggregate and/or the ice-like structure of the water in the pores. In the latter case the dyeing process is accelerated by the increase of D_{true}. Other so-called 'carrier actions' can often result from the fact that the added accelerator, being adsorbed competitively (cf. (11.5)), blocks some of the adsorption sites and thus makes possible more rapid penetration into the fibres (smaller $c_a^\varphi \to$ larger D_{app}: cf. (11.67)). In order to achieve more uniform and complete penetration of the fibre, 'retarders', which reduce the speed of dyeing, are used. Their action depends on the fact that the dyeing speed is determined by the desorption kinetics of the retarder ($\tau_D \ll \tau_d$). Analogous effects also play a large part in the kinetics of combination dyeings; here, one dye can act as a carrier (diffusion control) or as retarder (desorption control) of the other.

In principle, with the aid of this model, the dyeing of yarn can be treated in a similar way to the dyeing of a single fibre. The transport of dye between the separate threads of a yarn can be regarded as an outer diffusion process which is accompanied by an immobilization reaction (diffusion within the separate fibres). The ratio of the relaxation times of the two processes is closely connected with the problems of the penetration, migration and levelling[26,32] of dyes on textile materials.

The above survey of the fundamentals of dyeing kinetics by means of a simple statistical mechanical model is only indicative, because under conditions of practical dyeing the exact analysis of all simultaneous processes is not possible. For instance, it is quite customary to change the temperature, the salt concentration or the dyestuff concentration of the dye bath during the dyeing process.

The temperature dependence of the sorption rate in polymer masses is often calculated from the WLF equation (11.75) (Williams, Landel and Ferry[41,42]) for temperatures above the glass transition temperature:

$$\log \frac{D_{T_g}}{D_T} = -[C_1^g(T - T_g)]/(C_2^g + T - T_g) \qquad (11.75)$$

T_g = glass transition temperature

T = temperature of dyeing

D_{T_g} = experimentally determined diffusivity at T_g

D_T = experimentally determined diffusivity at T

C_1^g and C_2^g = constants of the dyeing system

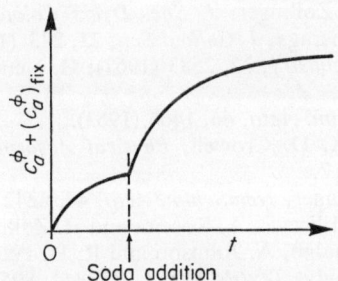

Figure 11.10 Dye-uptake curves of reactive dyestuffs

This relation may only be used as an approximate description of the experimental results. One obvious reason for this is the difficulty in determining meaningful diffusion coefficients. For example, under certain experimental conditions, the experimentally measured 'diffusion coefficient' according to the normally used method (slope of sorption curve) is time-dependent.

Finally, a typical adsorption curve for reactive dyes is depicted in Figure 11.10. The analysis of the fixation of reactive dyes is one of the most complicated problems in dyeing kinetics; in addition to diffusion and fixation processes, competing reactions (e.g. with water), competing adsorptions (reactive and hydrolysed dye), and the influence of preequilibria of aggregation and of tautomerism on the rate of fixation and the selectivity of the competing reactions[25] are important.

With reactive dyeing of wool, Figure 11.10, the kinetics become even more complex because the substrate possesses many chemically non-identical reaction sites.[34]

The fundamentals of dye application discussed in sections 11.3.1 and 11.3.2 apply not only to conventional aqueous dyeing processes but also to dyeing from the gas phase[35] or from solvents.[36]

11.4. LITERATURE

1. H. Raab, in *Ullmann's Encyclopädie der technischen Chemie*, 3rd edition, Urban and Schwarzenberg, München, 1962, Vol. 13, 806.
2. H. Gärtner, *J. Oil Colour Chemists' Assoc.*, **46**, 13 (1963).
3. D. Patterson, *Pigments*, Elsevier, Amsterdam, 1967.
4. J. S. Moll, *J. Soc. Dyers Colourists*, **76**, 141 (1960); C. B. Stevens, *Paint, Oil and Colour*, **147**, 80 (1965).
4a. F. Jones, *Rev. Progr. Color.*, **1**, 15 (1970).
5. B. Milicevic, *Chimia*, **16**, 29 (1962).
6. D. G. Miller, *Chem. Reviews*, **60**, 15 (1961); see also ref. 5.
7. J. O. Hirschfelder, C. F. Curtiss and R. B. Bird, *Molecular Theory of Gases and Liquids*, John Wiley, New York, 1954.
8. H. S. Frank and M. W. Evans, *J. chem. Physics*, **13**, 507 (1945); G. Nemethy and H. A. Scheraga, ibid., **36**, 3382, 3401 (1962); *J. physic. Chem.*, **66**, 1773 (1962); W. Kauzmann, *Advances Protein Chem.*, **14**, 1 (1959); R. P. Marchi and H. Eyring, *J. physic. Chem.*, **68**, 221 (1964); W. Luck, *Fortschr. chem. Forsch.*, **4**, 653 (1964); J. Lee Kavanau, *Water and Solute-Water Interactions*, Holden-Day Inc., San Francisco, 1964; H. Zollinger, *J. Soc. Dyers Colourists*, **81**, 345 (1965); D. C. Poland and H. A. Scheraga, *J. Colloid Sci.*, **21**, 273 (1966); C. Hansch and S. M. Anderson, *J. org. Chemistry*, **32**, 2583 (1967); G. Nemethy, *Angew. Chem.*, **79**, 260 (1967).
9. B. Milicevic, *Helv. chim. Acta*, **46**, 1466 (1963).
10. D. M. Young and A. D. Crowell, *Physical Adsorption of Gases*, Butterworths, London, 1962, Chap. 7.
11. G. Back and H. Zollinger, *Helv. chim. Acta*, **44**, 2242 (1958), **42**, 1526, 1539, 1553 (1959); G. Back, B. Milicevic, A. Roseira and H. Zollinger, *Melliand Textilber.*, **42**, 73 (1961); M. Greenhalgh, A. Johnson and R. H. Peters, *J. Soc. Dyers Colourists*, **78**, 315 (1962); H. Brady, *Textile Res. J.*, **35**, 844, 895 (1965).
12. B. G. Ferrini and H. Zollinger, *Helv. chim. Acta*, **50**, 897 (1967).
13. B. Milicevic, *Textilveredlung*, **3**, 607 (1968).
13a. R. McGregor and P. W. Harris, *J. Appl. Polymer Sci.*, **14**, 513 (1970).
14. Th. de Donder, *L'Affinité*, Gauthier-Villars, Paris, 1927.
15. T. Vickerstaff, *The Physical Chemistry of Dyeing*, Oliver and Boyd, London, 1954.
16. V. Hladik, *Fundamentals of Dyeing Theory*, SNTL/ALFA, Prague, 1968 (in Czech).
17. G. A. Gilbert and E. K. Rideal, *Proc. Roy. Soc. [London]*, **182A**, 335 (1944).
18. A. N. Derbyshire, *Trans. Faraday Soc.*, **51**, 909 (1955); *Discuss. Faraday Soc.*, **16**, 140 (1954).
19. T. Iijima and M. Sekido, *Sen-i Gakkaishi*, **18**, 153 (1962).
20. B. Milicevic and R. McGregor, *Helv. chim. Acta*, **49**, 1302, 1319, 2098 (1966).
21. P. B. Weisz and H. Zollinger, *Melliand Textilber.*, **48**, 70 (1967).
22. J. Crank, *Mathematics of Diffusion*, Clarendon Press, London, 1964.
22a. P. F. Dismore and W. O. Statton, *J. Polymer Sci. C*, **13**, 133 (1966); P. Predecki and W. O. Statton, *J. Appl. Physics*, **37**, 4053 (1966); W. O. Statton, 'Setting of synthetic fibres' in J. W. S. Hearle and L. W. C. Miles, *Setting of Fibres and Fabrics*, Manchester, 1970.
23. G. Kanig, *Kolloid-Z., Z. Polymere*, **190**, 1 (1963).
24. H. H. Sumner, *J. Soc. Dyers Colourists*, **76**, 672 (1960).
25. P. Rys, *Textilveredlung*, **2**, 95 (1967); P. Rys and H. Zollinger, *Helv. chim. Acta*, **49**, 749, 761 (1966); A. Datyner, P. Rys and H. Zollinger, ibid., **49**, 755 (1966).

26. P. B. Weisz, *Trans. Faraday Soc.*, **63,** 1801 (1967); P. B. Weisz and J. S. Hicks, ibid., **63,** 1807 (1967); P. B. Weisz and H. Zollinger, ibid., **63,** 1815 (1967); **64,** 1693 (1968); also ref. 21; P. B. Weisz, *Ph.D.Thesis* ETH, Zürich, 1966. See also: W. R. Vieth and K. J. Sladek; *J. Colloid Sci.*, **20,** 1014 (1965); W. R. Vieth, P. M. Tam and A. S. Michaelis, ibid., **22,** 360 (1966); W. R. Vieth, C. S. Frangoulis and J. A. Rionda ibid., **22,** 454 (1966).
27. P. B. Weisz and R. D. Goodwin, *J. Catalysis*, **2,** 397 (1963).
28. H. Sand, *Z. Elektrochem.*, **69,** 333 (1965); *Kolloid-Z., Z. Polymere*, **218,** 30, 124 (1967)
29. E. Atherton, D. A. Downey and R. H. Peters, *Textile Res. J.*, **25,** 977 (1955); E. Atherton and R. H. Peters, ibid., **26,** 497 (1956); see also: R. H. Peters, 'The kinetics of dyeing' in J. Crank and G. S. Park, *Diffusion in Polymers*, Academic Press, London/New York, 1968, 343; T. Iijima, T. M. A. Hossain, H. Maeda and Z. Morita, *Polym. Letters*, **5,** 1069 (1967).
30. E. Merian, *Textile Res. J.*, **36,** 612 (1966); see also ref. 26.
31. H. Zollinger, *J. Soc. Dyers Colourists*, **81,** 345 (1965); E. Golling, *Z. angew. Physik*, **14,** 717 (1962).
32. E. Merian, J. Carbonell, U. Lerch and V. Sanahuja, *J. Soc. Dyers Colourists*, **79,** 505 (1963).
33. G. Valk, M.-L. Kehren and E. Loers, *Textilveredlung*, **4,** 46 (1969); see also B. S. Sprague, *J. Polymer Sci.*, **20,** 159 (1967).
34. J. Shore, *J. Soc. Dyers Colourists*, **84,** 408, 413, 545 (1968); **85,** 14 (1969).
35. F. Jones and R. Seddon, *Textile Res. J.*, **34,** 373 (1964), **35,** 334 (1965); F. Jones and J. Kraska, *J. Soc. Dyers Colourists*, **82,** 333 (1966).
36. B. Milicevic, *Textilveredlung*, **4,** 213 (1969); *Rev. Progr. Color.*, **1,** 49 (1970).
37. Nobuhiko Kuroki, *The Theoretical Chemistry of Dyeing*, Maki Shoten, Tokyo, 1966 (in Japanese).
38. P. V. Danckwerts, *Trans. Faraday Soc.*, **46,** 300 (1950); **47,** 1014 (1951).
39. J. H. Petropoulos and P. P. Roussis, *J. Chem. Physics*, **47,** 1491, 1496 (1967), **48,** 4619 (1968), **50,** 3951 (1969).
40. F. T. Lindstrom, R. Hague and W. R. Coshow, *J. physic. Chem.*, **74,** 495 (1970).
41. M. L. Williams, R. F. Landel and J. D. Ferry, *J. Am. chem. Soc.*, **77,** 3701 (1955).
42. S. Rosenbaum, *J. Polymer Sci.*, **A3,** 1949 (1965); T. M. A. Hossain, H. Maeda, T. Iijima and Z. Morita, *Polymer Letters*, **5,** 1069 (1967); T. Iijima and S. Ikeda, *Angew. Makromol. Chem.*, **14,** 177 (1970).

Subject Index

Absorption of light,
and chemical constitution, 5ff., 23
visual colorimetry, 15ff.
Acedianthrone, 147, 156
Aceto acetanilides, as azo coupling components, 29, 42, 46, 52, 160
1-Acetylaminoanthraquinone dyes, 153ff., 160
Acid dyes (see also anionic dyes), 4, 48ff., 58, 72, 73, 107, 131, 149
Acridine dyes, 100, 101, 103, 155
Acridone dyes, 138, 139, 154, 155
Activity-gradient model of dyeing, 180ff.
Addition–elimination mechanism, 24, 34
Adsorption of dyes,
equilibria, 164ff.
fluorescent dyes, 109
general, 160
indigoid dyes, 115
selective, 169
sites for, 165ff., 180
to silver halides, 84ff.
to ZnO, 58
Affinity,
definition by Donder, 171
definition by Gilbert and Rideal, 172
of dye to substrate, 48, 49, 52, 63, 153
standard, 63, 153, 170ff.
Agfacolor process, 88
Aggregation of dyes (see association)
Alcian Blue 8 GX, 98
Algol Yellow GC, 154
Alizarin, 1, 2, 125, 126, 127, 129, 151
Alizarine Blue SWR, 128
Alizarine Bordeaux B, 128
Alizarine Cyanine Green G Extra, 149
Alizarine Yellow R, 54
Alkali fastness, 49, 50
Allopolar isomerism, 76, 77
Aluminium complexes, 151

Amidinium ions, 74, 75
Amido Yellow E, 72
Amination, aminolysis, 24, 34
1-Amino-4-bromo-anthraquinone-4-sulphonic acid (see bromamine acid)
4-Aminodiphenylamine derivatives, 52
1-Amino-iminoisoindolenine, 98
1,8,3,6-Amino-naphthol-disulphonic acid (H-acid), 50, 61, 66, 93
Amino-naphthol-sulphonic acids,
2,8,6-, (γ-acid), 46, 59
2,5,7-, (I-acid), 60, 61
o-Aminophenols,
diazotization (see o-Diazophenols)
in sulphur dyes, 110
2-(4'-Aminophenyl-)-5-methylbenzthiazole, 58
Aniline,
derivatives, as azo coupling components, 51
p-nitro, diazotized, 44, 51, 54
titration of derivatives with $NaNO_2$, 45
Aniline Black, 107
'Aniline dyes', 101
Anionic dyes, 3, 48ff., 58, 74, 75, 107, 163, 166
Annulenes, 92
Anthanthrone, 141, 152, 155, 160
1,2,9,10-Anthradiquinone, 127
Anthranilic acid, 56
Anthranol derivatives, 158
Anthrapyrimidine dyes, 147, 151, 156
Anthraquinone,
reduction, 157
synthesis, 126
Anthraquinone derivatives,
electronic spectrum, 124ff.
synthesis, 126ff.
Anthraquinone dyes, 3, 13, 34, 35, 48, 51, 69, 114, 124ff., 160, 161

Anthrimides, 133, 139, 140, 154
Anthrone derivatives, 84
Anti-sensitization, 85
Aromaticity, 93, 146
Artisil Blue GLF, 150
Arylazenium dyes, 74, 100ff.
Arylcarbonium dyes, 74, 99ff.
Arylogue, definition, 99
Aryltetracarboxylic acids, 147ff., 156
Aryne, 24, 33ff.
Association of dyes, 69, 76, 85, 153, 164, 166ff., 182, 183
Astra Blue Base 6 GLL, 97
Astraphloxine FF, 83
Astrazon Red 6 B, 84
Atebrin, 101
Auramin O, 106
Auxochrome (see electron donor)
Aza annulene dyes, 3, 37, 92ff.
Aza compounds, definition, 74, 83
Azamethine dyes, 44, 99, 100
Azidinium salts, 44
Azine dyes, 100, 104
Azobenzene, 11, 12, 27, 42
Azo coupling reaction,
    application, 42, 45ff., 161
    mechanism, 28ff., 31ff., 45ff.
    oxydative, 43ff., 83, 88, 104
Azo dyes, 1, 12, 18, 42ff., 147, 153, 154, 160, 161
Azo Geranine 2 G, 50
Azoic dyes, 4, 52ff., 58, 160, 179

Barbituric acid, 52
Basazol dyes, 69
Base catalysis, specific *vs.* general, 31ff., 36, 46ff., 64, 66, 67, 131
Basic dyes (see cationic dyes)
Bathochromic shift, 11, 12, 58, 76, 82, 115, 153
Benzanthrone, 143, 145, 147
Benzene, electronic states, 6, 7, 93
Benzidine, bisdiazotized, 46, 59
Benzo Fast Copper Red GGL, 61
Benzoquinone, 34, 35
Benzothiazol derivatives,
    in azo dyes, 43, 51
    in polymethine dyes, 74, 76, 77, 82, 83
    in sulphur dyes, 110ff.
Benzyne (see aryne)
Bindschedler's Green, 107

Bisdiazotization (see also benzidine), 46, 57
Bismarck Brown, 1
Bohn–Schmidt reaction, 128
Brighteners, optical, 4
Bromamine acid, 69, 131, 133, 134, 138, 149, 151

Calcobound dyes, 69
Caledon Blue XRN (see indanthrone)
Caledon Jade Green XN, 156
Capri Blue GON, 106
Carbazol dyes, 112, 138ff., 154
Carbolan Blue B, 149
Carbolan Green G, 149
Carbolan Violet 2 R, 149
Carbomethine dyes (see trimethine dyes)
Carbonyl dyes, definition, 114
o-Carboxy-o'-hydroxyazo dyes, 58
β-Carotene, 84
Carotenoid dyes, 84
Carrier, 168, 169, 182
Cationic dyes, 3, 74, 83, 106, 149, 163, 166
Cavalite dyes, 64ff.
Celliton Fast Blue Green B, 150
Celliton Fast Yellow 7 GF, 84
Celliton Scarlet B, 51
Cellulose acetate fibres, 149
Cellulose fibres,
    dyeing, 2, 3, 51, 54, 58ff., 63, 107, 110, 112, 153ff., 156ff., 179
    hydroxyl groups, 63, 179
    structure, 164
Chelates (see metal complexes)
Chelation effect, 39
Chlorantine Fast Red 5 BRL, 61
Chlorantine Fast Turquoise Blue BLL, 97
Chlorantine Light Green BLL, 61
2-Chlorobenzthiazole dyes, 64ff.
Chlorophyll, 37, 92, 94
Chromaticity diagram, 16, 17, 19
Chrome Violet, 107
Chroming of dyes on fibres, 54, 58, 160
Chromium complexes, 54ff., 97, 107, 151
Chromophore (see electron acceptor)
Chrysophenine G, 59
Ciba Blue 4 B, 2
Cibacron Brilliant Red 3 BL, 66
Cibacron dyes, 64ff.
Cibalan Brilliant dyes, 2
Cibanone Blue RSN (see indanthrone)

Cibanone Red G, 154
CIE system, 15ff.
Classification of dyes, 3
Cobalt complexes, 54ff., 94, 95, 98, 151
Colorimetry,
  Lambert–Beer, 9
  visual, 15ff.
Colour (see absorption of light, colorimetry, colour vision)
Colour development (see colour photography)
Colour Index, 4
C.I. Acid Black 58, 55, 56
C.I. Acid Blues,
  138, 149
  158, 55
C.I. Acid Greens,
  1, 73
  16, 107
  25, 149
  50, 107
C.I. Acid Oranges,
  3, 72
  7, 49
  20, 49
C.I. Acid Reds,
  1, 50
  87, 106
C.I. Acid Violet 51, 149
C.I. Acid Yellow 73, 106
C.I. Basic Blues,
  7, 106
  9, 106
C.I. Basic Green 4, 100, 103, 106
C.I. Basic Reds,
  2, 107
  12, 83
C.I. Basic Violets,
  3, 100, 102, 106
  7, 84
  10, 107
C.I. Basic Yellow 2, 106
C.I. Direct Black 78, 62
C.I. Direct Blues,
  71, 61
  86, 97
C.I. Direct Green 26, 61
C.I. Direct Reds,
  1, 59
  80, 61
  180, 61

C.I. Direct Violet 82, 61
C.I. Direct Yellow 12, 59
C.I. Disperse Blues,
  7, 150
  20, 150
C.I. Disperse Red 1, 51
C.I. Disperse Yellows
  13, 150
  31, 84
C.I. Ingrain Blues,
  1, 98
  7, 98
C.I. Mordant Black 11, 55, 56
C.I. Mordant Orange 1, 54
C.I. Mordant Violets,
  26, 128
  39, 107
C.I. Pigment Blues,
  14, 97
  15, 97
C.I. Pigment Greens,
  7, 97
  8, 73
C.I. Reactive Blue 6, 150
C.I. Solubilized Vat Blue 1, 156
C.I. Solvent Blues,
  22, 107
  51, 97
C.I. Solvent Yellow 2, 2, 11
C.I. Sulphur Blacks,
  1, 113
  12, 113
C.I. Sulphur Blues,
  9, 112
  13, 112
C.I. Sulphur Green 14, 97
C.I. Sulphur Yellow 4, 112
C.I. Vat Black 27, 154
C.I. Vat Blues,
  4 (see indanthrone)
  29, 98
  33, 155
C.I. Vat Brown 3, 154
C.I. Vat Greens,
  1, 156
  3, 155
  8, 154
C.I. Vat Oranges,
  3, 155
  7, 156

C.I. Vat Reds,
  13, 156
  15, 156
  23, 156
C.I. Vat Violet 14, 155
C.I. Vat Yellows
  2, 154
  31, 156
Colour photography, 29, 34, 85ff., 101, 104
Colour vision (perception), physiology, 21, 84, 85
Complementary colour, 17
Complexes, 1:1 and 1:2, definition (see also metal complexes, metal complex dyes), 55
Conformation, of azo dyes at azo-N, 58
Congo Red, 59
Coordination numbers, 38, 55ff.
Copper complexes, 38, 54, 61ff., 69, 92, 94ff.
Coprantine Violet BLL, 61
Cotton (see cellulose fibres)
Coulomb attraction, 163
Coupling component, 28, 42, 45ff., 51, 52
Cromophthal dyes, 2, 54
Crystal Violet, 100, 102, 106
Cyanidine, 88
Cyanine dyes, 74, 75ff., 84ff., 115
Cyanuric chloride,
  in direct dyes, 61
  in reactive dyes, 64ff.

DABCO, 65
Danckwerts equation, 182
Dehydrophthalocyanine, 92, 94
Dehydrothiotoluidine, 58, 111
Delocalization energy, 7
Desorption (see adsorption)
Development, of colour film, 86
Diamine Fast Red F, 59
1,5-Diamino-4,8-naphthoquinone, 146, 150
4,4'-Diaminostilbene-2,2'-disulphonic acid, 59
o-Dianisidine, 59
Diarylcarbonium dyes, 99ff., 106
1,4-Diazabicyclo-[2,2,2]-octane, 65
Diazoamino compounds, 28, 53
Diazo components,
  aromatic, 28ff., 42, 44, 45ff., 49, 54
  heterocyclic, 43, 51ff.
Diazocyanides, 27

Diazo decomposition reactions, 46
Diazo group transfer, 42, 44
Diazohydroxides, 26, 27
Diazonium ions (salts),
  as an acid, 26, 27
  as diazo component, 27ff., 42
  mechanism of formation, 26
  reactions, 24, 26, 27, 28, 33
  reaction with reducing reagents, 42
  stabilization, 37, 45, 53
o-Diazophenols, as diazo components, 46, 47, 48
Diazosulphonates, 27, 43, 53
Diazotates,
  acid-base equilibria, 27
  cis-$trans$ isomerism, 27ff., 46
  $trans$, for Rapid Fast dyes, 53
Diazotization,
  application, 44ff., 52ff.
  by gas fumes, 150
  mechanism, 26, 45
Dibenzanthrone dyes, 141ff., 152, 156
Dibenzpyrenequinones ($cis/trans$), 141, 142, 152, 155
Dicarbomethine dyes (see pentamethine dyes)
Dichloroindigo, isomers, 118
4,5-Dichloropyridazone dyes, 64ff.
2,3-Dichloroquinoxaline dyes, 64ff.
Dichlorotriazine dyes, 64ff.
2,3-Dicyan-1,4-dithiacyclohexane(2,3), 97
Diffusion of dyes,
  diffusion coefficients, 175ff.
  general, 160, 162, 174ff.
  in Polaroid process, 89ff.
  shell-progressive, 180
  Stokes-Einstein relation, 178
1,2-Dihydronaphthalene, 36
1,2-Dihydroxyanthraquinone (see Alizarin)
o,o'-Dihydroxyazo dyes, 39, 40, 54ff.
N-Dimethylaniline, 12
1,1-Dimethylhydrazine, 65
N,N'-Dimethylindigo, 118
Dimethyl Yellow, 11, 12
2,4-Dinitro-1-naphthol, 72
Dioxazine dyes, 107, 160
1,4-Diphenylaminoanthraquinone derivatives, 114, 149, 150, 151
Diphenylmethane dyes (see diarylcarbonium dyes)

Subject Index

Direct dyes, 4, 54, 58ff., 164
Disazo dyes,
 general, 42, 46, 48, 51, 52, 58ff.
 primary/secondary, definition, 58
Disperse dyes, 2, 3, 4, 13, 51ff., 84, 146, 149, 150ff., 164
Distribution principle of auxochromes, 13, 114, 118, 125, 146, 150
Dithionite, sodium, 152
Donnan membrane theory, 170
Dow phenol process, 24, 33
Drimarene dyes, 64ff.
Duranol Brilliant Yellow 6 G, 150
Dyeing, methods of (see also cellulose acetate fibres, cellulose fibres, equilibria, kinetics, polyacrylonitrile fibres, polyamide fibres, polypropylene, protein fibres), 160ff.

Ektachrome process, 88
Electron acceptor, 11, 12, 13, 150
Electron donor, 7, 12, 13, 150
Electronic spectra (see absorption of light)
Electrophotography, 58, 84
Electrostatic forces, 163
Elimination–addition mechanism, 24
Enols, as azo coupling components, 29, 45, 46
Entropy,
 of dye aggregation, 167
 of dyeing equilibria, 174
 of hydrophobic interactions, 164
Eosine, 106
Equilibria, of dyeing process, 162, 163ff.
Eriochrome Black T, 55, 56
Excited states, 6, 9, 10, 11, 155
Exhaustion dyeing, 160
Extinction, 9

Fanal dyes, 161
Fast Black Base K, 52
Fast Blue Z, 107
Fast colour salts, 52
Ferri complex dyes, 72, 92, 95
Fibre-damage, by vat dyes, 155
Fick's laws,
 1st, 175ff.
 2nd, 177ff., 180
Fischer base (see 1,3,3-trimethyl-2-methylene-indoline)

Fixation, in dyeing process (see immobilization)
Flavanthrone, 129, 142, 155, 160
Fluorescein, 106
Fluorescence, 11, 83, 106, 108
Food dyes, 4, 84
Formazan dyes, 85
Franck–Condon principle, 10
Free electron model, 8, 76
Freundlich isotherm (see isotherm)
Friedel–Crafts reactions, 31, 121, 126, 138, 141
Fuchsine, 1

γ-acid (see 2,8,6-amino-naphthol-sulphonic acid)
Gibbs–Helmholtz equation, 174
Glass transition temperature and dyeing, 183

H-acid (see 1,8,3,6-amino-naphthol-disulphonic acid)
Haemin, 37, 92, 94
Halogenations, 24, 31
Heat of dyeing, 174
Hemicyanine dyes, 74, 75, 84
Hemioxonol dyes, 76, 77, 84, 88
Henry's law, 171
Hering theory, 21
Herz reaction, 120ff.
Hexakisazo dyes, 61
Hildebrand activity coefficient, 170
HMO method, 6, 7, 92, 115
Hückel rule, 7, 92, 93, 146, 147
Hydrogen bond, 7, 125, 146, 150, 164
Hydride ion acceptors, 35, 37, 89, 123, 142
Hydron Blue, 112
Hydrophobic interactions, 164
Hydroquinone, quinone-hydroquinone system, 34ff., 158
Hydrosulfite (see dithionite)
Hydroxyanthraquinones, 126
Hydroxyazo-quinonehydrazone tautomerism, 43, 50, 51, 55, 58
2-Hydroxy-3-naphthoic acid,
 anilide, 52, 58, 160, 179
 other derivatives, 54, 160
Hydroxynitroso-quinoxime tautomerism, 72ff.
1-(2'-Hydroxyphenylazo)-2-naphthol, 55
Hypsochromic shift, 11, 76, 115, 153

I-acid derivatives (see 2,5,7-aminonaphthol-sulphonic acid)
Immedial Black V, 113
Immedial Brilliant Blue CLB, 112
Immedial Yellow GG, 112
Immobilization, of dye in dyeing process, 63, 174ff., 179ff.
Indamine dyes, 100
Indanthrene Blue RS (see indanthrone)
Indanthrene Bordeaux RR, 156
Indanthrene Brilliant Blue 4 G, 98
Indanthrene Brilliant Orange GR, 156
Indanthrene Brilliant Orange RK, 155
Indanthrene Brown R, 154
Indanthrene Khaki GG, 154
Indanthrene Olive Green B, 155
Indanthrene Olive R, 154
Indanthrene Red Brown 3 R, 156
Indanthrene Red GG, 156
Indanthrene Red Violet RRK, 155
Indanthrene Rubine R, 156
Indanthrene Turquoise Blue 3 GK, 155
Indanthrene Yellow 2 GF, 154
Indanthrene Yellow GK, 155
Indanthrene Yellow 4 GK, 156
Indanthrene Yellow 5 GK, 154
Indanthrone, 2, 125, 129, 138, 152, 155, 160
Indicators, dyes as, 107ff.
Indigo, 1, 2, 13, 114ff., 146, 152, 153, 156
Indigofera tinctoria, 1, 115
Indigosol dyes, 156ff.
Indigosol O, 156
Indirubin dyes, 115
Indoaniline dyes, 100
Indophenol dyes, 100
Indoxyl, 119ff.
Ingrain dyes, 4
Ink, dyes for, 106
Intermediate, definition, 24, 30
Ion exchange mechanism, in dyeing, 48, 166ff., 178
Irgalan dyes, 2
Irgalan Grey BL, 55
Isatin derivatives, 119ff.
Isatis tinctoria, 115
Isoindigo derivatives, 115
Isotherm,
 distribution, sorption, 162, 163ff.
 Freundlich, 164ff.

Langmuir, 163, 164ff., 178
Nernst, 164ff.
Isotope effect, kinetic, hydrogen, 31ff.
Isoviolanthrone, 141, 142, 145, 156

Kinetics, of dyeing processes, 162, 163, 164, 174ff.
Knudsen diffusion, 175
Kodachrome process, 88
Kodacolor process, 88
Kubelka–Munk equation, 20

Lakes (see pigment dyes)
Lambert–Beer law, 9
Lanasol dyes, 68
Langmuir isotherm (see isotherm)
LCAO-MO method, 6, 95, 125
Leather dyes, 4, 48
Leuco compounds, 103, 104, 112, 123, 124, 133, 137, 143, 151ff., 156ff., 164
Leuco sulphuric acid ester dyes, 156ff.
Levafix dyes, 64ff.
Levelling, in dyeing, 182
Lewis activity coefficient, 171
Light fastness, 150
Luminosity, 16ff.

MacAdam ellipses, 17
Malachite Green, 100, 103, 106
Mauveine, 1, 101, 107
Maxilon dyes, 83
Mercurochrome, 101
Merocyanine dyes, 76, 79, 82, 84, 85, 88, 100, 115
Mesomeric structures, 7, 8, 74, 98
Metal complex dyes, 2, 37, 51, 54ff., 61ff., 69, 72ff., 92ff., 107, 151, 160, 161
Metal complexes (see also metal complex dyes), 24, 37ff., 54ff.
Metameric colours, 15
Methine dyes, 29, 44, 74ff.
Methylene Blue, 106
Michler's hydrol, 102ff.
Michler's ketone, 102
Migration in dyeing, 182
Milling dyes, 49
MIM method, 8, 12, 118, 125
Molecular orbitals (see also orbitals), 6ff.
Monastral Fast Blue B, 97

Monastral Fast Blue G, 97
Monastral Fast Green G, 97
Monoazahemioxonols, 83, 88
Monoazo dyes, 42, 45ff., 48ff., 66, 67, 69, 83
Monochlorotriazine dyes, 64ff.
Monomethine dyes, 74
Mordant dyes (see also metal complex dyes), 4, 37, 151
Morse curve, 10, 11
Munsell colours and terms, 19ff.

Naphthalene Green V, 107
Naphthalene Orange I, 49, 50, 59, 164
Naphthalene Orange G, 49, 164
1,4,5,8-Naphthalenetetracarboxylic acid, 148, 156, 160
Naphthazarin dyes, 146
2-Naphthol-3-carboxylic acid (see 2-hydroxy-3-naphthoic acid)
2,6,8-Naphthol-disulphonic acid, 31, 32
Naphthol Green B, 73
Naphthols,
　1-, 43, 46, 49, 50
　2-, 43, 46, 49, 56, 57, 61
Naphthol-sulphonic acids,
　1,3-, 31, 32, 47
　1,4-, 31, 32
　2,6-, 28
　2,8-, 31, 32
Naphthylamine derivatives (see also aminonaphthol-sulphonic acids and -disulphonic acids), 49, 61
Naphtol AS, 52ff., 58, 160, 179
Naphtol AS-G, 52
Natural dyes, 1, 4, 37, 77, 84, 92, 115
Neolan Blue 2 G, 55
Neolan dyes, 2
Nernst isotherm (see isotherm)
Neutrocyanine dyes, 76
Neutrogene dyes, 53
Nickel complexes, 39, 94, 95, 151
Nitration, 24, 25, 26, 31
Nitric acid, equilibria, 25
Nitro dyes, 72ff.
Nitronium ion (see nitryl ion)
Nitroso dyes, 72ff.
Nitrosyl bromide, 45
Nitrosyl sulphuric acid, 44

Nitrous acid,
　equilibria, 26
　in diazotization, 26ff., 44ff.
Nitryl ion, 25

o/p-ratio, in azo coupling reactions, 48, 49
Optical density (see extinction)
Orange I and II (see Naphthalene Orange I and G)
Orbitals, 5, 94
Orientation,
　in azo coupling, 33, 46, 47ff., 49
　in indigoid dyes, 117
Oscillator strength, 10
Over-dyeing, 167, 168
Over-sensitization, 85
Oxazine dyes, 100, 104
Oxidative coupling reactions (see azo coupling reactions)
Oxindigo, 117
Oxonol dyes, 75, 79, 84, 85, 100

π-complexes, 33, 36
π-electrons, 7, 8, 9, 11, 12, 92
Paper dyes, 48
Particle size, of pigments, 161
p-electrons, 7, 8, 94
Pentamethine dyes, 74, 76ff.
Pentazamethine dyes, 44
Perception, of colour, 15, 17, 19, 20, 21
3,4,9,10-Perylenetetracarboxylic acid, 148, 156, 160
Phenazones, in sulphur dyes, 110
Phenolphthalein, 100, 103, 107ff.
Phenols,
　acid–base equilibrium, 28
　as azo coupling components, 28ff., 45ff.
　cyano, 12
　in sulphur dyes, 110ff.
　nitro, 12
　synthesis, 24, 33, 34
Phenothiazones, in sulphur dyes, 110
Phenoxazones, in sulphur dyes, 110
Phenylazo-naphthols,
　tautomerism, 43, 51
　4,1-, 43, 50
Phenylene Blue, 107
Phenylenediamine,
　derivatives for colour photography, 87ff.

derivatives in anthraquinone dyes, 124, 133, 136
p-, 35
Phenylglycine derivatives, in (thio)indigo synthesis, 119ff.
Phenylogue, definition, 99, 100
Phosgene, 61, 101
Phosphorescence, 11
Photography (see also colour photography), electrophotography, 58, 84
silver bromide, 84ff.
Phthalein dyes, 100, 103
Phthalocyanine dyes, 2, 3, 37, 69, 92ff., 161
Phthalogen Blue Black IVM, 98
Phthalogen dyes, 98
Pigment dyeing, 160ff.
Pigment dyes, 2, 4, 48, 53ff., 73, 97, 106, 107, 137, 146, 151, 155, 160ff.
Pigment Green B, 73
Pigment printing, 161
Platinum complexes, 95
Polaroid process, 89ff.
Polyacrylonitrile fibres, dyeing, 83, 84, 106, 149, 163, 166, 179
Polyamide fibres, dyeing, 3, 48, 54, 63, 67, 151, 163, 167
Polyazacyanine dyes, 75, 82, 83, 99
Polyazo dyes, 42, 46, 47, 54, 58
Polyene dyes, 84, 115
Polymethine dyes,
properties, 83ff.
structure, 74ff., 100
synthesis, 79ff.
Polypropylene, dyeing, 107, 151
Ponsol Blue GZ (see indanthrone)
Pore model of dyeing, 165, 174ff.
Porosity, 176
Porphin, 92
PPP method, 8, 13, 115
Preequilibria, 24, 25, 33, 45, 46
Premetallized dyes, definition, 56
Primazine dyes, 64ff., 67
Primulin base, 111
Procinyl Blue RS, 151
Procinyl dyes, 67
Procion dyes, 2, 64ff.
Production of dyes, industrial, 1, 2, 23, 42, 115
Protein fibres, dyeing, 2, 3, 48, 51, 54, 59, 63, 68, 106, 107, 149, 153, 163, 166

Purity (CIE), 17ff., 58
Purple, ancient (6,6′-dibromoindigo), 1
Pyranthrone, 141, 142, 155
Pyrazolanthrone dyes, 147, 152, 156
Pyrazolones,
for azo dyes, 29, 42, 46, 49, 52, 67, 160
for methine dyes, 76, 89
3-methyl-1-phenyl-5-pyrazolone, acid–base equilibrium, 29, 46
Pyridine,
catalysis of azo couplings, 33, 46ff.
in indigosol synthesis, 156
Pyrogene Indigo GW, 112
Pyrylium dyes, 77

Quercetin, 77
Quinacridone, 2, 146, 160
Quinalizarin, 128
Quinhydrone, 35
Quinoline derivatives, as azo coupling components, 52
Quinone, quinone-hydroquinone system, 34
Quinone hydrazones (see hydroxyazo-quinonehydrazone tautomerism)
Quinoneimines and -diimines, 35, 43, 100, 103, 104, 105, 107

Radical mechanisms,
aromatic substitution, 24
quinone-hydroquinone system, 35
Raoult law, 171
Rapidazol dyes, 53
Rapid Fast dyes, 53
Rapidogen dyes, 53
Reactive dyes,
constitution, 51, 63ff., 151
dyeing, 160, 179ff., 183
for cellulose fibres, 2, 62ff., 97, 98
for wool, 2, 4, 62ff., 151, 183
hydrolysis, 63ff., 183
reactive groups, 3, 62ff., 97, 98
tautomerism, 183
Reactolan dyes, 64ff.
Reactone dyes, 64ff.
Reatex dyes, 64ff.
Redox reactions,
indicators, 107ff.
mechanism, 34ff.
vat dyes, 152, 157, 158

Subject Index

Reference state, in thermodynamics, 172
Relaxation time, 177, 178, 179
Remalan dyes, 2, 67
Remazol dyes, 67ff.
Remazol Golden Yellow G, 67
Resonance theory (see mesomeric structures, valence bond method)
Retarder, in dyeing, 182
Rhodamine B, 107
Rhodanine, 82
Rongalite C, 152
RW-acid, 60

σ-complex, 24, 33
Safranine T Extra, 107
Salicylic acid, 39, 54, 58, 59, 61, 97, 107
Salt effects,
  in azo coupling reactions, 46
  in dyeing, 168, 169
Sandocryl dyes, 149
Scarlet acid, 60
Scatchard plot, 169
SCF method, 8
Scholl reaction, 34, 35ff., 142
Schrödinger equation, 5, 6, 8
Screening dyes, 85
Selenindigo, 117
Semiconductors,
  Fe-phthalocyanine, 95
  ZnO, 58, 84
Semiquinone, 35
Sensitivity, of human eye, 15ff., 84
Sensitization,
  of AgO, 77, 84ff.
  of semiconductor properties of ZnO, 58
S_E2 reactions (see substitution, aromatic, electrophilic)
Silk (see protein fibres)
Sirius Light Blue BRR, 61
Solvatochromy, 8, 76
Solvent dyeing, 183
Solvent dyes, 4
Sorbate, definition, 163
Sorbent, definition, 163
Sorption (see adsorption)
Spin dyeing, pigments for, 161
Stabilization of diazonium salts (see diazonium ions)
Steady state,
  in diffusion, 176

intermediates, 31
Stereochemistry, of metal complexes, 39ff., 58
Steric hindrance
  in azo coupling reactions, 47
  in dyeing processes, 168
Stilbene dyes, 59
Stokes–Einstein relation, 178
Streptocyanine dyes, 74, 75, 100
Substantive dyes (see direct dyes)
Substantivity (see affinity)
Substitution, aromatic,
  electrophilic, 24ff., 29ff., 87, 101, 122, 124, 128, 138, 145
  homolytic, 24
  nucleophilic, 24, 33ff., 101, 129ff., 145
Sulphamic acid, in diazotization, 44
Sulphophthalein dyes, 100
Sulphur Black T, 113
Sulphur dyes, 2, 4, 58, 97, 110ff.

Technicolor process, 90
5,5',8,8'-Tetrabromothioindigo, 160
Tetrachloro-1,2-quinone, 36
Tetracyanoethane in colour photography, 87
Tetrakisazo dyes, 42, 61
Tetrazaporphin, 92, 97
Tetrazotization, 46
Thermodynamics,
  irreversible, 163, 174
  of dyeing equilibria, 170ff.
  of dyeing kinetics, 174
Thianthrene dyes, 112, 113
Thiazine dyes, 100, 104, 105, 161
Thioindigo, 2, 117ff., 153, 160
Thioindoxyl, 119ff.
Thionol Ultra Green B, 97
Thiophenol derivatives, synthesis, 120
Thioxanthen dyes, 100
o-Tolidine, 59
Tortuosity, 176
Transitions, electron, 6, 9, 10, 11
Transition state, definition, 24, 30
Triarylcarbonium dyes, 9, 29, 48, 85, 99ff., 106ff., 161
Triaxines, 1,3,5-trichloro- (see cyanuric chloride)
2,4,6-Trihalogenopyrimidine dyes, 64ff.
Trimethine dyes, 74

1,3,3-Trimethyl-2-methylene-indoline, 52, 78, 79
Triphenylmethane dyes (see triarylcarbonium dyes)
Trisazo dyes, 42, 58, 59, 93
Tristimulus values, 16
Tropylium ion, 7
Turkey Red, 1, 2, 151

Ullmann reactions, 129ff., 142
Urea,
   de-associating effect, 47
   in diazotization, 44

Valence bond method, 8
Van der Waals' forces, 163
Variamin Blue Base B, 52
Vat acid, 152
Vat dyes, 4, 13, 34, 58, 98, 112, 114ff., 141ff., 151ff., 160, 164, 179

Verofix dyes, 64ff.
Victoria Pure Blue BO, 106
Violanthrone, 141, 144, 145, 156, 160
Vitamin A, 84
Vitamin B_{12}, 37

Water, structure, 164, 182
Wavelength, dominant, 17ff.
Weber–Fechner law, 19
Williams–Landel–Ferry equation, 183
Witt's theory, 5, 12
Wool (see protein fibres)
Wool Green BS, 107

Xanthene dyes, 100, 103, 106, 108

Young–Helmholtz–Maxwell theory, 21

Zambesi Black V, 62
Zinc complexes,
   in electrophotography, 58, 84
   phthalocyanine, 95

DATE DUE